Single-Electron Devices and Circuits in Silicon

Single-Electron Devices and Circuits in Silicon

Zahid Ali Khan Durrani

Imperial College, UK

Imperial College Press

Published by

Imperial College Press
57 Shelton Street
Covent Garden
London WC2H 9HE

Distributed by

World Scientific Publishing Co. Pte. Ltd.
5 Toh Tuck Link, Singapore 596224
USA office: 27 Warren Street, Suite 401-402, Hackensack, NJ 07601
UK office: 57 Shelton Street, Covent Garden, London WC2H 9HE

British Library Cataloguing-in-Publication Data
A catalogue record for this book is available from the British Library.

SINGLE-ELECTRON DEVICES AND CIRCUITS IN SILICON
Copyright © 2010 by Imperial College Press

All rights reserved. This book, or parts thereof, may not be reproduced in any form or by any means, electronic or mechanical, including photocopying, recording or any information storage and retrieval system now known or to be invented, without written permission from the Publisher.

For photocopying of material in this volume, please pay a copying fee through the Copyright Clearance Center, Inc., 222 Rosewood Drive, Danvers, MA 01923, USA. In this case permission to photocopy is not required from the publisher.

ISBN-13 978-1-84816-413-0
ISBN-10 1-84816-413-0

Printed in Singapore.

To my wife, Olivia, and to my parents,
Khalid and Rahila

Preface

Single-electron devices provide a means to control electronic charge at the level of one electron, by means of the single-electron charging or 'Coulomb blockade' effect. These devices operate by controlling the transfer of charge across tunnel barriers onto nanometre-scale conducting regions or 'islands'. In such a process, the energy needed to charge an island with even one electron can be large enough to influence the tunnelling process. This energy, the 'single-electron charging energy', must be overcome to allow current to flow across the island, preventing current flow at low applied voltage and temperature.

The possibility that the single-electron charging energy of a nanostructure could influence the tunnelling of even one electron onto the nanostructure was identified as early as the 1950s. In 1951, C.J. Gorter proposed that the observed increase in resistance in thin, granular metal films at low electric fields and temperatures was associated with the need to overcome the single-electron charging energy of the nanometre-scale grains in the film. In the mid-1980s, K.K. Likharev and co-workers predicted, in great detail, effects relating to single-electron charging in nanometre-scale tunnel junctions. By this stage, advances in nanofabrication techniques had led to the ability to fabricate well-defined, nanometre-scale, islands and tunnel junctions. In 1987, this led to the first demonstration, at low temperature, of a *designed* single-electron device, the single-electron transistor (SET) of Fulton and Dolan.

Over the following two decades, a wide variety of single-electron devices, in many material systems, were demonstrated. These included numerous different types of SETs, single-electron memory devices, single-electron logic circuits, and devices for the controlled transfer of charge packets formed by single or small numbers of electrons. Furthermore, there were many demonstrations of the room-temperature operation of single-electron devices, particularly in silicon semiconductor material.

The great interest in single-electron devices has been driven in part by the potential of these devices for applications in large-scale integrated (LSI) circuits. In comparison with conventional semiconductor devices, devices such as the SET and the single-electron memory cell are inherently nanometre-scale, and tend to improve in performance when scaled down in size. Furthermore, these devices possess the advantages of very low power consumption, associated with the small amounts of charge they use, and control over any statistical fluctuations in this charge.

This book discusses the design, fabrication and electrical characterization of single-electron devices and circuits in silicon. We concentrate on single-electron devices in silicon, as these are of particular interest for LSI circuit applications. Single-electron devices in metals, and in other semiconductor systems such as GaAs/AlGaAs heterostructures, are discussed only when necessary to understand the operation of particular device types, or when device implementations in silicon are limited. This book considers the physics of single-electron charging effects. This is followed by a review of the fabrication and operation of SETs in crystalline and nanocrystalline silicon materials. Single-electron memory devices are then discussed, where the stored 'bits' are defined by single, or at most a few, electrons. We then consider few-electron charge transfer devices, such as single-electron pumps and turnstiles, where small numbers of electrons can be transferred through the device using radio frequency signals. Finally, we discuss single-electron logic circuits. Throughout this book, we follow an approach where the various types of single-electron devices are reviewed first, and then explained in more detail using examples from the author's research.

The author's research work would not have been possible without the collaboration and support of many colleagues. Firstly, the author would like to acknowledge the help and support of Prof. Haroon Ahmed at the Microelectronics Research Centre, University of Cambridge, UK. The major part of the author's research was conducted in collaboration with

Prof. Ahmed, and would not have been possible otherwise. The author would also like to acknowledge the great support of, and collaborations with, many others at the Microelectronics Research Centre, especially Dr John Cleaver, Dr Andrew Irvine, Dr David Hasko, Dr Yong Tsong Tan, Dr Toshio Kamiya, Dr Mohammed Khalafalla, Dr Jin He and Dr Aftab Rafiq (now at Tokyo Institute of Technology, Japan). This research was also conducted in collaboration with Hitachi Cambridge Laboratory, and in this regard, the author would like to acknowledge extremely valuable collaborations with, and the help and support of, Prof. Hiroshi Mizuta (now Professor of Nanoelectronics at the University of Southampton, UK), Prof. Kazuo Nakazato (now at the University of Nagoya, Japan) and Dr David Williams. Furthermore, a large part of the author's research has been possible only through longstanding and very productive collaborations with Prof. Shunri Oda at Tokyo Institute of Technology, Japan, and with Prof. William Milne at the Department of Engineering, University of Cambridge, UK. The author would also like to acknowledge the help and support of Prof. Richard Syms, Prof. Mino Green and Dr Kristel Fobelets, at the Department of Electrical and Electronic Engineering, Imperial College, London, UK, and the support of many other colleagues and friends at the Microelectronics Research Centre, Cambridge, Hitachi Cambridge Laboratory and the Department of Engineering, University of Cambridge. The author would like to acknowledge the financial support of various research programmes. These include the FASEM project (part of the ESPRIT programme of the European Union), the 'Single-electron logic' project, funded by EPSRC, UK, and the CREST and SORST programmes of Japan Science and Technology Corp., Japan.

Finally, the author would like to acknowledge the invaluable support and encouragement of his wife, Olivia Kesselring-Durrani, his parents, Khalid and Rahila Durrani, and his mother-in-law, Ursula Kesselring. Their support has made the writing of this book a far less arduous task.

Zahid A.K. Durrani

Contents

Preface	vii
1. Introduction	1
1.1 Single-Electron Effects	5
1.2 Early Observations of Single-Electron Effects	8
1.3 Basic Single-Electron Devices	12
1.3.1 Single-electron transistor.	13
1.3.2 Single-electron box	14
1.3.3 Multiple-tunnel junction	15
1.4 Scope of This Book	17
1.4.1 Introduction to subsequent chapters	18
2. Single-Electron Charging Effects	22
2.1 Introduction	22
2.2 Single Tunnel Junction	25
2.3 The Single-Electron Box.	28
2.3.1 The 'critical charge'.	29
2.3.2 Electrostatic energy changes	31
2.3.3 Energy diagram for the single-electron box.	36
2.4 The Single-Electron Transistor	37
2.4.1 Electrostatic energy changes	39
2.4.2 Tunnelling rates.	43
2.4.3 Offset charge	48
2.4.4 Calculation of I-V characteristics.	49
2.4.5 The Coulomb staircase	52
2.4.6 Energy band diagrams	56
2.5 Quantum Dots	59
2.5.1 Coulomb oscillations in quantum dots.	62
2.6 The Multiple-Tunnel Junction	64

3. Single-Electron Transistors in Silicon 72
   3.1 Early Observations 72
   3.2 SETs in Crystalline Silicon 75
       3.2.1 SETs with lithographically defined islands 75
             3.2.1.1 Etched islands. 76
             3.2.1.2 Pattern-dependant oxidation 82
       3.2.2 SETs using MOSFET structures 85
       3.2.3 Crystalline silicon nanowire SETs 87
       3.2.4 Room temperature Coulomb oscillations with large peak-to-valley ratio 91
       3.2.5 Fabrication and characterization of nanowire SETs 93
             3.2.5.1 Fabrication 94
             3.2.5.2 Electrical characterization 97
   3.3 Single-Electron Transistors in Nanocrystalline Silicon 101
       3.3.1 Conduction in continuous nanocrystalline silicon films 103
       3.3.2 Nanocrystalline silicon nanowire SETs 107
       3.3.3 Point-contact nc-Si SET: Room temperature operation...... 111
       3.3.4 'Grain-boundary' engineering 116
       3.3.5 SETs using discrete silicon nanocrystals 117
       3.3.6 Comparison with crystalline silicon SETs. 118
       3.3.7 Electron coupling effects in nanocrystalline silicon 119
             3.3.7.1 Electrostatic coupling effects 121
             3.3.7.2 Electron wavefunction coupling effects. 122
   3.4 Single-Electron Effects in Grown Si Nanowires and Nanochains... 124

4. Single-Electron Memory 130
   4.1 Introduction 130
       4.1.1 Multiple-tunnel junction memory. 132
   4.2 MTJ Memories in Silicon. 136
       4.2.1 The single-electron detector. 138
   4.3 Single- and Few-Electron Memories with Floating Gates 140
   4.4 Large-scale Integrated Single-Electron Memory in Nanocrystalline Silicon 146
   4.5 Few-Electron Memory with Integrated SET/MOSFET 149
       4.5.1 Silicon nanowire SETs for L-SEM application 151
             4.5.1.1 Nanowire SET in crystalline silicon. 153
             4.5.1.2 Nanowire SET in polycrystalline silicon 154
             4.5.1.3 Potential for mass fabrication 155
       4.5.2 Single-gate L-SEM 156
             4.5.2.1 Single-gate L-SEM fabrication and characterization 159
       4.5.3 Split-gate L-SEM 163

4.5.4	L-SEM 3 × 3 cell array	167
	4.5.4.1 Memory cell selection	168
	4.5.4.2 Temperature dependence of memory cell characteristics	172

5. Few-Electron Transfer Devices — 174

5.1 Introduction 174
5.2 Single-Electron Turnstiles and Pumps 175
    5.2.1 Single-electron turnstile 176
    5.2.2 Single-electron pump 179
    5.2.3 Single-electron pump and turnstile using a semiconductor quantum dot 181
5.3 Few-Electron Devices using MTJs 184
    5.3.1 Operation of single r.f. signal MTJ electron pump 186
    5.3.2 Single r.f. signal MTJ electron pumps in GaAs 191
    5.3.3 Single r.f. signal MTJ electron pumps in silicon 193
        5.3.3.1 Device fabrication and experimental Characteristics 195
    5.3.4 MTJ electron pump with multi-phase r.f. signals 198
5.4 Single-Electron Transfer Devices in Silicon 200
    5.4.1 Single-electron transfer using a CCD 201
    5.4.2 SET/MOSFET single-electron pump and turnstile 203
5.5 Metrological Applications 206

6. Single-Electron Logic Circuits — 209

6.1 Introduction 209
6.2 Voltage State Logic 210
    6.2.1 SET inverter with resistive load 211
    6.2.2 Complementary SET inverter 213
    6.2.3 Complementary SET NAND and NOR gates 219
    6.2.4 Programmable SET logic 222
    6.2.5 Logic using SETs with multiple input terminals 225
    6.2.6 Effect of offset charge 227
6.3 Charge State Logic 228
    6.3.1 Binary decision diagram logic 229
        6.3.1.1 Binary decision diagram logic: Basic logic gates 231
        6.3.1.2 Implementation of BDD logic circuits in GaAs 237
    6.3.2 Implementation of BDD logic circuits in silicon 245
        6.3.2.1 Two-way BDD switch using silicon nanowire SETs 246

        6.3.2.2 Extension to a 'universal' three-way switch 250
   6.4 Quantum Cellular Automaton Circuits...................... 252
   6.5 Single-Electron Parametron 258

Bibliography   261

Index   281

Chapter 1

Introduction

Single-electron devices provide a means to precisely control the charging of a small conducting region at the level of one electron. These devices operate using the Coulomb blockade or single-electron charging effect (Devoret and Grabert, 1992; Likharev, 1999), where the energy associated with the addition or subtraction of one electron from a nanometre-scale electrode controls the electrical characteristics of the device. In comparison with conventional semiconductor devices, single-electron devices such as the single-electron transistor (SET) (Fulton and Dolan, 1987) and the single-electron memory cell (Nakazato *et al.*, 1993) are inherently nanometre-scale and highly scalable. Furthermore, these devices possess the advantages of ultra-low power consumption, associated with the very small amounts of charge they use, and immunity from statistical fluctuations in the charge (Nakazato *et al.*, 1993; Yano *et al.*, 1999). This has led to great interest in these devices for future LSI circuit applications.

Since the 1970s, the speed and performance of LSI circuits has improved dramatically, associated with a continuous reduction in the size of semiconductor devices. The minimum feature size in an integrated circuit has reduced from >1 µm in 1970 to ~50 nm in 2008, and it is expected that by 2011, it may be possible to define features smaller than ~20 nm (International Technology Roadmap for Semiconductors, 2007). At present, the physical gate length in high-performance metal-oxide semiconductor field-effect transistors (MOSFETs) may be as small as ~35 nm (Mistry *et al.*, 2007). At least with respect to the channel length, we may regard LSI MOSFETs as *nanoscale* devices, i.e. with dimensions in the range 1–100 nm.

Conventional LSI circuit devices such as the silicon MOSFET, dynamic random access memory (DRAM) cells, and FLASH memory cells, operate using 'bits' defined by large numbers of electrons. For example, a DRAM cell uses ~100,000 electrons to define the '1' bit (Yano et al., 1999). Similarly, silicon MOSFETs operate with large numbers of electrons, each electron behaving simply as a carrier of charge. We may therefore regard conventional LSI device operation as 'classical' and simply approximate the influence of quantum mechanical effects. Quantum mechanical effects in a heavily-scaled MOSFET include the gate leakage current associated with the tunnelling of electrons across thin gate oxide layers, corrections to the threshold voltage caused by energy quantization in the potential well of the inversion layer, and the possibility that a fraction of the channel current is associated with the ballistic transport of electrons (Taur and Ning, 1998).

With the reduction in MOSFET size deep into the nanoscale, short-channel effects (SCE) associated with the breakdown of the long-channel approximation in a MOSFET lead to an increase in the device current and a reduction in the threshold voltage (Taur and Ning, 1998). The increasing significance of SCE with reduced device size leads to the degradation of gate control over the channel current, an increase in the device 'off' current, and an increase in the static power consumption. This problem is most serious for high-speed devices, where the channel length is scaled most heavily. In high performance, sub-50 nm channel length MOSFETs, the drain source 'off' current $I_{sd,leak}$ is already ~30 nA/μm (ITRS, 2007). It is expected that, if the channel length reduces to ~10 nm by 2013, $I_{sd,leak}$ will increase to ~100 nA/μm or greater. The situation is somewhat better for larger (65 nm channel length), lower speed, low operating power MOSFETs, where $I_{sd,leak}$ ~1 nA/μm. However, this is also expected to increase, reaching ~10 nA/μm by 2013 (16 nm channel length). For even larger, low standby power devices, at present $I_{sd,leak}$ ~10 pA/μm (75 nm channel length) and is expected to increase to ~100 pA/μm by 2013 (18 nm channel length). The increase in $I_{sd,leak}$ complicates and may even limit device scaling, especially for high performance devices. Even in the best case, it is likely that complex system-level solutions may be needed to limit excessive power consumption (ITRS, 2007).

Other issues also arise as device size reduces into the nanoscale. Process tolerances may lead to variations in the structure of nominally identical devices, e.g. variations in gate-oxide thickness or channel length. These effects cause greater variation in the characteristics of a nanoscale device. Furthermore, charge fluctuations between different devices, e.g. changes in the charge in the channel or fluctuations in the doping concentration in nominally similar devices, become increasingly significant and may lead to variation in the electrical characteristics between different devices (Ferry and Goodnick, 1997). As an example, a nanoscale region 10 nm × 50 nm × 1 µm in size, doped n-type at a concentration of $10^{17}/cm^3$, would possess on average only 5 dopant atoms. Statistical '$\sqrt{n}$' variations in this number of dopant atoms are clearly unacceptable, and higher doping levels would be needed.

Since the 1980s, the ability to fabricate nanoscale devices has led to a large body of work on novel semiconductor devices directly using quantum mechanical effects for their operation (Ferry and Goodnick, 1997). These include *quantum dots* in vertical and planar III-V heterostructure material (Reed *et al.*, 1988; Kouwenhoven *et al.*, 1991), where quantum confinement of electrons in a potential well in one or more dimensions influences the electrical characteristics, ballistic transport devices such as quantum point-contacts (van Wees *et al.*, 1988, Wharam *et al.*, 1988), and single-electron devices. The latter allow control over charge at the level of *one* electron. Initially, the great interest in these devices was driven by investigations of *mesoscopic* physics, i.e. the physics of structures larger than the atomic scale, but smaller than the macroscopic scale, where Boltzmann transport theory would apply (Ferry and Goodnick, 1997).

Early, relatively large (~100 nm) quantum effect devices required cryogenic temperatures to work. While this is not an issue for experiments investigating the physics in these devices, clearly it prevents the practical application of such a device in most cases. However, the potential of single-electron devices such as the SET, and the single-electron 'box' for integrated circuit applications was identified even in the earliest investigations (Likharev, 1988). Clearly, if charge storage on a nanostructure could be controlled at the one electron level, then the number of electrons necessary to define a single bit could be dramatically reduced, leading to a large reduction in the power consumption of the device (Yano *et al.*, 1999). Control of charge at the one electron level would also eliminate any statistical, '$\sqrt{n}$' variations in the electron number n, removing the effect of these variations in large numbers of devices (Nakazato *et al.*, 1993). Furthermore, if a quantum device was fabricated using conventional LSI-compatible materials and fabrication techniques, e.g. if the device was fabricated in silicon, then LSI fabrication technology would be available for the circuit integration of the device. In addition, quantum devices often operate at very small current levels. Amplification of these levels would then be necessary for interfacing with other electronic devices, on or off a chip. Compatibility with conventional LSI technology implies that MOSFETs could be used to provide this interface, allowing quantum devices to communicate with the 'outside world'.

Devices operating using quantum mechanical principles can often *improve* in performance with reduction in device size (Yano *et al.*, 1999). For example, in a single-electron device, a reduction in the size of the device to ~10 nm leads to an increase in the maximum operating temperature of the device to room temperature (Takahashi *et al.*, 1995). Similarly, the energy level separation in a quantum dot increases with a reduction in the dot size (Saitoh and Hiramoto, 2002). Furthermore, the voltage levels in these devices also increase, to values closer to those used in conventional MOSFETs. In contrast, MOSFETs become increasingly complex in structure as the device size is reduced and effects such as SCE or tunnelling across the gate oxide tend to degrade the electrical characteristics. The development of practical single-electron devices allows us to envisage low power, highly integrated circuits using devices only ~10 nm in scale, operating with ultra-small charge packets containing, at most, a few electrons.

1.1 Single-Electron Effects

Single-electron charging effects provide a means to control the charge on a small, nanometre-scale conducting region at the level of one electron (Devoret and Grabert, 1992; Likharev, 1999). Consider a conducting region fabricated in either a metal or a doped semiconductor, with dimensions ~100 nm or less. The size of this region is such that it forms a *nanostructure*. Furthermore, consider that the conducting region, or *island*, is near two other conducting regions, or electrodes, and that an insulating material lies in the gaps (Fig. 1.1[a]). If the widths of the gaps are small, each ~10 nm or less, then a voltage difference applied across the electrodes can transfer electrons on to, and off the island by quantum mechanical tunnelling. The gaps then form tunnel barriers, with an associated energy barrier (Fig. 1.1[b]). Such a system is called a single-island, double tunnel junction.

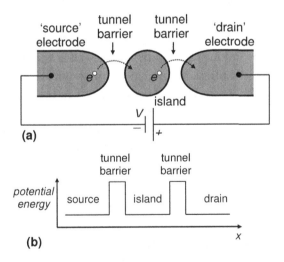

Fig. 1.1 The single-island, double tunnel junction system. (a) Schematic diagram, where the island and electrode regions (shaded) are formed by conducting materials. The gaps between these regions form tunnel barriers. (b) The potential energy across the system, at zero bias.

Electrons can tunnel onto the island only if the charging energy associated with adding the electrons to the island is overcome (Devoret and Grabert, 1992). If the total capacitance of the island is C, then the charging energy of a *single* electron added to the island is given by $E_c = e^2/2C$. This is referred to as the *single-electron charging energy*. It is essential that this energy is overcome before an electron can tunnel onto the island. With a nanometre-scale island ~100 nm or less in size, it is easily possible that $C \sim 10^{-16}$ F or less (Fulton and Dolan, 1987). This implies that the single-electron charging energy $E_c \geq 0.8$ meV. At cryogenic temperatures, e.g. at the temperature of liquid Helium, $T = 4.2$ K, the thermal energy $k_B T = 0.36$ meV, where k_B is Boltzmann's constant. We therefore have a system where $E_c > k_B T$ and the thermal energy may not be enough to allow an electron to tunnel onto the island. If we were to reduce the island size further, reducing C and increasing E_c, then E_c may be $>> k_B T$ and the likelihood of an electron tunnelling onto the island because of the thermal energy it possesses becomes very small. If $C \sim 10^{-18}$ F, a possibility if the island size is ~10 nm (Takahashi *et al.*, 1995), then $E_c \sim 80$ meV, greater than $k_B T = 26$ meV at room temperature, $T = 300$ K. This implies that, even at room temperature, the thermal energy is insufficient for an electron to tunnel onto the island.

Assuming that $E_c >> k_B T$, we now attempt to transfer electrons onto the island by applying a voltage V_{ds} across the two electrodes. We shall now refer to the electrodes as the drain and source electrodes, a bias being applied to the drain electrode and the source electrode being considered as 'ground'. The circuit diagram of the system is shown in Fig. 1.2(a). Here, the tunnel junctions have capacitances C_1 and C_2, and resistances R_1 and R_2, and the total island capacitance is $C = C_1 + C_2$. The voltage V_{ds} creates a potential difference between the intermediate island and each of the electrodes. For a small positive value of V_{ds}, the potential difference V_1 between the higher energy, source electrode and the island is small enough such that $E_c = e^2/2C$ is not overcome. Therefore, electrons on this electrode cannot tunnel onto the island. However, if V_{ds} is increased such that E_c is overcome, an electron can tunnel onto the island. This electron can then tunnel off the island to the lower energy, positively biased drain electrode, and a current begins to flow across the system. A further electron can then tunnel immediately onto the island, repeating the process. On average, the island charge increases by one electron. In a similar manner, for negative values of V_{ds} large enough such that potential difference between the drain and the island is large enough to overcome E_c, electrons can tunnel from the drain to the source, with current flowing in the opposite direction. For low values of V_{ds}, electrons cannot overcome the single-electron charging energy and cannot tunnel onto the island, and a current cannot flow. This is called the *Coulomb blockade* effect, and leads to the *I-V* characteristics shown in Fig. 1.2(b). Ideally, at 0 K, a zero current region or *Coulomb gap* exists for $|V_{ds}| < |V_c|$, where V_c is the *Coulomb blockade voltage*. At small finite temperatures, a small thermally activated current exists within the Coulomb gap.

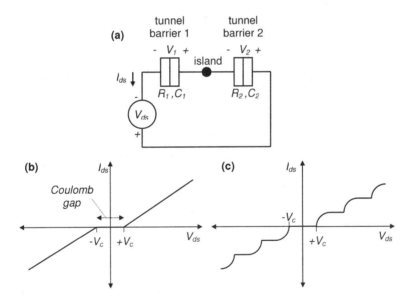

Fig. 1.2 (a) Circuit diagram of the single-island, double tunnel junction system. (b) Coulomb blockade *I-V* characteristics. If the tunnel barriers are similar, the current increases in magnitude linearly outside the Coulomb gap. (c) Coulomb staircase *I-V* characteristics. If the tunnel barriers are very dissimilar, the current increases in magnitude in a stepwise manner.

We now increase V_{ds} beyond the Coulomb gap edge $+V_c$. When V_{ds} reaches a value such that the charging energy corresponding to two electrons $2E_c = e^2/C$ is overcome, a second electron can persist on the island, and on average, two extra electrons exist on the island. With further increases in V_{ds}, the number of electrons charging the island increases one-by-one. If the tunnelling rates across the two tunnel junctions are similar, the current increases linearly outside the Coulomb gap. However, if the tunnelling rates are very different, then the current rises in a non-linear, stepwise manner, creating a *Coulomb staircase* in the *I-V* characteristic (Fig. 1.2[c]). Each successive step in the characteristics corresponds to an increase in the number of electrons on the island by one.

A further restriction on single-electron charging effects is imposed by the nature of the tunnel barriers that quasi-isolate the island. If the tunnel barriers are too conductive then charging electrons cannot be localized on the island. Quantum mechanically, for single-electron charging to occur, the electron wavefunction must not extend strongly across the tunnel barriers into the electrodes. Localization of electrons on the island is possible if the tunnel barrier resistance R_t is greater than the resistance quantum $R_k = h/e^2 \approx 25.8$ kΩ, where h is Plank's constant, i.e. $R_t >> R_k$ (Devoret and Grabert, 1992). In practice, this usually requires that R_t is at least ~$10R_k$. Furthermore, it is necessary for the tunnel barrier height E_t $>> k_BT$, in order to prevent thermally activated current over the potential barrier. The relatively large value of R_t in single-electron devices implies that these are high resistance devices in comparison with MOSFETs.

1.2 Early Observations of Single-Electron Effects

The possibility that the single-electron charging energy of a nanoscale island influences subsequent tunnelling events to the island was identified as early as the 1950s. In 1951, C.J. Gorter proposed that the observed increase in resistance of thin, granular metal films at low electric field and temperature was associated with the need to overcome the energy required to transfer an electron from one grain to another (Gorter, 1951). Figure 1.3(a) shows a scanning electron micrograph of a thin metal film of Au, nominally ~5 nm thick, evaporated on GaAs. The film is strongly granular in nature, with a range of grain sizes. The smallest grains are only ~10 nm, small enough for significant single-electron charging effects. In further measurements, Neugebauer and Webb (Neugebauer and Webb, 1962) explained the in-plane conductivity of various granular metal films with grains ~1 – 10 nm in size, at low temperature down to 77 K, using a model where electron transfer between the grains was determined by an activation energy associated with the single-electron charging energy.

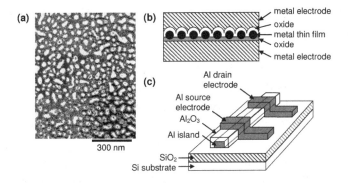

Fig. 1.3 (a) Scanning electron microscope (SEM) image of a granular Au film, average thickness 5 nm, evaporated on a GaAs substrate. Single-electron effects can be observed in the in-plane conduction through such a film, associated with the Au nanoparticles. (b) Metal electrode/oxide/granular metal film/oxide/metal electrode structure. Single-electron effects caused by the nanoparticles in the granular metal film can be observed in I-V or C-V characteristics measured across the two electrodes. (c) The Al/Al$_2$O$_3$ single-electron transistor, demonstrated by Fulton and Dolan in 1987.

These early works were extended by measurements of single-electron charging in thin metal films sandwiched between top and bottom contacts. Figure 1.3(b) shows a schematic diagram of such a structure. A thin oxide layer is grown or sputtered onto a metal contact. This is followed by the evaporation of a thin metal film on top, forming a layer of metal nanoparticles. A thin oxide layer is then grown or sputtered on top of the granular metal film, and finally a metal contact is deposited on top. The thin oxide layers form tunnel barriers for I-V measurements across the structure. Alternatively, one of the oxide layers can be thicker, forming a capacitor. C-V measurements can be performed on such a structure, where electrons are transferred into the metal nanoparticles only across the thinner oxide layer by tunnelling. Giaever and Zeller, in measurements of tin nanoparticle films embedded within an insulating matrix, observed an increase in the film resistance at low voltages in both normal and superconducting mode, at temperatures down to 1.6 K (Giaever and Zeller, 1968; Zeller and Giaever, 1969). This was explained by considering the charging energy of a single electron on a nanoparticle. Lambe and Jaklavic (Lambe and Jaklavic, 1969), in C-V measurements at 4.2 K on capacitors formed by metal nanoparticles (a system very similar to the 'single-electron box' described in the next section), observed charge quantization on the nanoparticles. These works, and further quantitative analysis by Kulik and Shekhter (Kulik and Shekhter, 1975), led to a good understanding of single-electron effects in granular metal films by the 1970s. By the mid-1980s, Likharev and co-workers (Likharev and Zorin, 1985; Averin and Likharev, 1986) predicted theoretically in great detail the behaviour of a single nanoscale tunnel junction. This was followed by the observation of a Coulomb staircase in the I-V, and differential resistance dI/dV vs. V, characteristics of granular films with well-defined tunnel barriers (Kuzmin and Likharev, 1987; Barner and Ruggiero, 1987). These characteristics corresponded to the addition of electrons one-by-one onto the island of a double tunnel junction.

In the preceding experiments, the island was defined 'naturally' by the granular nature of a thin metal film. By the late 1980s, advances in nanofabrication techniques implied that it was possible to fabricate well-defined nanoscale islands and tunnel junctions. This led to the demonstration of the first SET, by Fulton and Dolan (Fulton and Dolan, 1987), with a Coulomb gap in the I-V characteristics, and oscillations in the Coulomb gap with gate voltage. Figure 1.3(c) shows a schematic diagram of an Al/Al$_2$O$_3$ SET, formed on a SiO$_2$ on Si substrate. The island is formed by an aluminium electrode, and aluminium oxide tunnel barriers are formed by oxidizing the island. Two further aluminium electrodes, deposited across the island, form source and drain contacts. The silicon substrate may be used as a gate electrode, coupled capacitively to the island. In the device of Fulton and Dolan, the island electrode was formed by a ~14 nm thick Al layer, with dimensions

~800 nm × 50 nm. The total island capacitance $C \sim 1$ fF and $E_c \sim 100$ μV, requiring measurements at milli-Kevin temperatures such that $E_c \gg k_BT$. The Coulomb gap in the device was ~1 mV.

It is possible to fabricate well-defined islands and tunnel barriers in the Al-Al$_2$O$_3$ system, and this system is widely used for metal single-electron devices. However, these devices typically operate at cryogenic temperatures, and raising the maximum temperature of single-electron effects requires far smaller islands. In the first observation of a Coulomb staircase at room temperature, Schönenberger et al. (Schönenberger et al., 1992) used a scanning tunnelling microscope to characterize a thin granular metal film with islands only ~1 nm size. Furthermore, even in the early stages of single-electron research, a large number of applications of single-electron devices, from quantum metrology, to sensitive electrometers, to transistors, memory and logic devices for integrated circuit applications, were identified (Likharev, 1988).

Single-electron effects in a semiconductor system were first observed in the conductance of a one-dimensional (1-D) channel of a silicon MOSFET at low temperature (Scott-Thomas et al., 1988, and the follow-up paper by Van Houten and Beenaker, 1989). Here, disorder in the channel potential isolated a segment of the channel between tunnel barriers, forming an island for single-electron charging. Small islands were also defined by lithography in the two-dimensional electron gas (2-DEG) layer in III-V heterostructures. In these devices, patterned surface gates were used to form the tunnel barriers, by depleting sections of the 2-DEG in AlGaAs/GaAs heterostructure material (Meirav et al., 1990; Kouwenhoven et al., 1991a). At low temperatures, quantum confinement effects occur simultaneously with single-electron effects in the island, forming a quantum dot.

A very large body of work now exists on quantum dots in 2-DEGs, driven by investigations of the physics of zero-dimensional systems (Kouwenhoven et al., 1997). Single-electron devices may also be fabricated in silicon-on-insulator (SOI) material, by patterning the top silicon layer using high-resolution electron beam lithography and etching (Ali and Ahmed, 1994). This provides a very flexible means to fabricate single-electrons devices with nanoscale islands, e.g. the first SET operating at room temperature was fabricated by Takahashi et al.

(Takahashi et al., 1995) in SOI material using electron beam lithography and oxidation. Here, the island size was only ~10 nm, the island capacitance $C \sim 10^{-18}$ F and $E_c \sim 0.1$ V.

There are a number of detailed, general reviews of single-electron devices. Early work on the subject has been reviewed by various authors (Likharev, 1988; Averin and Likharev, 1991; Schön and Zaiken, 1990; Devoret et al., 1992). The reader is also referred to the book *Single Charge Tunneling*, edited by Grabert and Devoret (Grabert and Devoret, 1992), covering the theoretical and experimental state-of-the-art in 1992 in great detail. There is a later general review by Likharev (Likharev, 1999). Semiconductor single-electron devices are reviewed by Meirav and Foxman (Meirav and Foxman, 1996) and devices in silicon by Takahashi et al. (Takahashi et al., 2002). Quantum dots are reviewed by Ashori et al. (Ashori et al., 1996), van Houten et al. (van Houten et al., 1993), and Kouwenhoven et al. (Kouwenhoven et al., 1997). Interacting or 'coupled' quantum dots are reviewed by van der Wiel et al. (van der Wiel et al., 2002). There are also various special issues of journals, focusing on single-electron and quantum dot devices (special issues of Z. Physik, 1991; Physica B, 1993; IEICE Transactions on Electronics, 1998). Furthermore, there are reviews for the more general reader, by Likharev and Claeson (Likharev and Claeson, 1992) and by Harmans (Harmans, 1992). Finally, a general review of electron transport in nanostructures, including single-electron devices, is given by Ferry and Goodnick (Ferry and Goodnick, 1997).

1.3 Basic Single-Electron Devices

We now introduce three basic single-electron devices, the single-electron transistor (SET) (Fulton and Dolan, 1987), the single-electron box (Lafarge et al., 1991) and the multiple-tunnel junction (MTJ) (Delsing, 1992; Nakazato et al., 1992). These three devices form the most common single-electron systems, and are the basis of more complex single-electron circuits such as single-electron memory (e.g. Yano et al., 1999; Irvine et al., 2000), single-electron logic circuits (Tsukagoshi et al., 1998) and single-electron electron transfer devices such as single-electron pumps and turnstiles (Geerligs *et al.*, 1990; Pothier *et al.*, 1991). Hybrid devices consisting of combinations of the three basic devices may also be fabricated, e.g. the MTJ/single-electron box hybrid (Nakazato *et al.*, 1993) can form a single-electron memory cell, a configuration which has been widely investigated.

1.3.1 Single-electron transistor

Adding a third 'gate' terminal, electrostatically coupled to the island of the simple double tunnel junction discussed earlier converts the system into an SET (Fulton and Dolan, 1987). The circuit diagram of the SET is shown in Fig. 1.4(a). Here, a capacitor C_g connects the island to the gate.

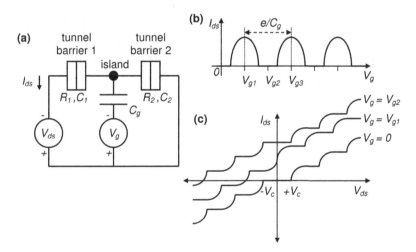

Fig. 1.4 (a) The SET. (b) Periodic single-electron oscillations in the I_{ds}-V_g characteristics, at a constant value of V_{ds}. (c) As V_g is varied, the edges of the Coulomb gap in the I_{ds}-V_{ds} characteristics vary periodically, e.g. from $\pm V_c$ at $V_g = 0$ V, to zero at $V_g = V_{g1}$, to $\pm V_c$ at $V_g = V_{g2}$. Here, for clarity, the three I_{ds}-V_{ds} characteristics are offset in I_{ds} from each other by equal amounts.

The gate voltage V_g may be used to control the Fermi level of the island, and overcome or impose a Coulomb blockade. This leads to the I_{ds}-V_{gs} characteristics shown schematically in Fig. 1.4(b), where at a constant value of V_{ds}, I_{ds} oscillates periodically. These characteristics, known as single-electron current oscillations, or Coulomb blockade oscillations, may be understood as follows. Applying a positive gate voltage V_g lowers the island Fermi energy, and at a value $V_g = V_{g1}$, the energy difference between the source and island caused by the single-electron charging energy is overcome. Electrons can then transfer from source to drain, across the island and a current is observed. We may view this as a reduction in the Coulomb gap to zero. Increasing V_{gs} lowers the island Fermi energy further. In Fig. 1.4(b), at $V_{gs} = V_{g2}$, the Coulomb blockade associated with a second electron on the island is yet to be overcome, and the current is very low. Increasing V_{gs} even further to V_{g3} overcomes the Coulomb blockade, and a second electron charges the island. Further increase in V_{gs} causes I_{ds} to oscillate periodically with a period e/C_g, each oscillation corresponding to a change in the electron number on the island by one. Furthermore, in the I_{ds}-V_{ds} characteristics (Fig. 1.4[c]), adjusting V_{gs} to a value corresponding to an oscillation peak leads to the Coulomb gap reducing to zero. Varying the gate voltage leads to a periodic oscillation in the Coulomb gap observed in the I_{ds}-V_{ds} characteristics. The SET can then be regarded as a simple switch, controlled by the gate voltage. For small values of V_{ds}, the SET is 'on' when the Coulomb blockade is zero and 'off' when a Coulomb blockade exists.

1.3.2 Single-electron box

Replacing one of the tunnel junctions of the single-island double tunnel junction by a capacitor forms the single-electron box (Lafarge *et al.*, 1991). The circuit diagram of the device is shown in Fig. 1.5(a), where the capacitor C_b replaces one of the tunnel junctions. C_b blocks DC current flow across the device, but charge may still be transferred onto or off the island from the source electrode, across the remaining tunnel junction. Applying a positive voltage V to the capacitor lowers the Fermi energy of the island relative to the Fermi energy of the source. When the Fermi energy of the island is lowered relative to the source by a value greater than the single-electron charging energy of the island, electrons can be transferred onto the island, charging C_b. Here, the single-electron energy is given by $E_c = e^2/2(C_1 + C_b)$. As before, it is necessary that $E_c > k_B T$, requiring that C_b is small. However, C_b must not behave as a tunnel capacitor, as electrons would then tunnel off the island. When the value of V_s is such that the Fermi energy difference between the source and the island just exceeds E_c, one electron transfers onto the island. As the Fermi energy difference exceeds $2E_c$, a second electron transfers onto the island and, in this way, electrons may be transferred onto the island one-by-one (Fig. 1.5[b]). Similarly, applying a negative voltage removes electrons from the island (assuming a metallic island) one-by-one.

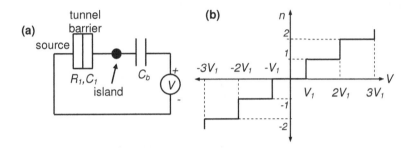

Fig. 1.5 (a) The single-electron box. (b) Electron number n on the island, as a function of applied voltage V.

We may contrast the single-electron box, where there is no DC current, with the double tunnel junction. In a double tunnel junction, it is also possible to charge the island with a precise number of electrons, but this is accompanied by a DC current across the device. The current is zero only within the Coulomb gap, when there are no extra electrons on the island.

1.3.3 Multiple-tunnel junction

The double tunnel junction and the single-electron box both contain only one island. It is also possible to observe single-electron charging effects with more than one island, connected to each other and to the electrodes by tunnel junctions. Such a system is referred to as a multiple-tunnel junction (MTJ) (Delsing, 1992; Nakazato *et al.*, 1992).

Figure 1.6(a) shows the circuit diagram of an MTJ with N tunnel junctions, and N-1 islands. In such a system, a much larger Coulomb gap is often observed. This may be considered, to an approximation, to be determined by the sum of the single-electron charging energies of the constituent islands (Delsing, 1992). The MTJ may be used as the basis of an SET, with a gate electrode coupled electrostatically to one or more islands (Fig. 1.6[b]). Furthermore, an MTJ version of the single-electron box is also possible (Fig. 1.6[c]), where the tunnel junction is replaced by an MTJ. Here, charge is transferred onto the capacitor C_b across the MTJ, if the source voltage V_s has a negative value high enough to overcome the Coulomb blockade of the MTJ.

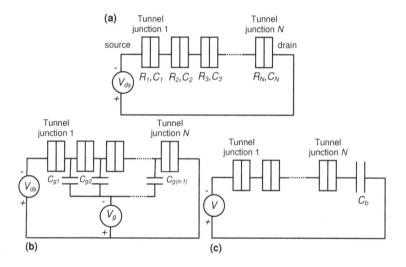

Fig. 1.6 (a) The MTJ. (b) The MTJ SET. (c) The MTJ single-electron box, or MTJ single-electron memory.

There are, however, significant differences between an electron box using an MTJ, and a simple single-electron box. Unlike a single-electron box, returning the source voltage V_s to zero after transferring electrons onto C_b does not remove the electrons charging C_b, as the MTJ remains within its Coulomb gap. A positive voltage is necessary to remove these electrons, leading to a 'memory' effect in the device. Furthermore, unless

C_b is very small, it will hold multiple electrons. However, it is easily possible to fabricate an MTJ electron box that holds only a few electrons (Nakazato *et al.*, 1993), forming a few-electron memory cell. It is also possible to reduce C_b to a value similar to the MTJ tunnel capacitances, leading to a true 'single-electron' memory cell (Stone and Ahmed, 2000). We note, however, that in all these cases, electrons are controlled by an MTJ in Coulomb blockade, i.e. by single-electron effects (Yano *et al.*, 1999). The MTJ memory cell may therefore be referred to as a 'Coulomb blockade' memory cell (Durrani *et al.*, 2000).

1.4 Scope of This Book

This book discusses the design, fabrication and electrical characterization of single-electron devices and circuits in silicon. We discuss the physics of single-electron charging effects in detail, and provide an introduction to quantum dots. The book then discusses the fabrication and operation of nanoscale SETs in various silicon-based material systems. Single-electron memory devices are reviewed, where the stored 'bits' are defined by a few tens of electrons at most. We then consider few-electron charge transfer circuits such as single-electron pumps and turnstiles, where packets of charge consisting of one, or at most, a few electrons are transferred through a circuit using radio-frequency (r.f.) signals. Finally, the application of single-electron devices for logic operations is discussed.

We follow an approach where various single-electron device designs in silicon are reviewed, and examples from the author's research are used to illustrate the fabrication process and the electrical characteristics in detail. The book focuses on silicon single-electron devices as these are compatible with LSI fabrication techniques, and are of particular interest for the development of future, highly-scaled LSI circuits. A wide range of single-electron devices and circuits have now been demonstrated in the laboratory, often with LSI applications in mind. However, as we have seen earlier in this chapter, single-electron devices have been investigated equally widely in many other material systems, e.g. the Al/Al_2O_3 metal island system, or in AlGaAs/GaAs heterostructure materials. However, these devices are perhaps of greater interest for investigations of the physics of nanostructures than for LSI circuit applications. We therefore consider devices in systems other than silicon only occasionally, usually when they pre-date equivalent devices in silicon and understanding their operation is essential to a discussion of later devices in silicon. For more information on single-electron devices in other systems, the reader is referred to the list of general review papers provided earlier in this chapter.

1.4.1 Introduction to subsequent chapters

The subsequent chapters of this book are organised as follows. Chapter 2, 'Single-Electron Charging Effects', describes the theoretical basis of single-electron effects. After a brief introduction, the basic requirements for single-electron charging in a nanoscale system are discussed. The chapter then introduces single-electron effects in a single tunnel junction, and the observation of single-electron tunnelling oscillations in the current through the junction. These oscillations, at a frequency $f_{SET} = I/e$, provide a means to link frequency, current and charge at a fundamental level. The chapter then discusses the single-electron box. In this more practical device, energy fluctuations in the environment are decoupled from the charging island. For simplicity, the device is analysed assuming a metallic island. The change in the total electrostatic energy of the island, when electrons are added or removed from the island, is calculated first. This is then used to calculate the electron tunnelling rate. The concept of the 'critical charge' is discussed, the extent of the Coulomb blockade region in the device is defined, and an energy band picture for the device is introduced. The chapter then discusses the SET, again assuming a metallic island. The electrostatic energy changes and electron tunnelling rates in the basic double tunnel junction, and in the SET, are calculated. The tunnelling rates are then used to derive an expression for the I-V characteristics of the SET. The Coulomb staircase and single-electron current oscillations in the SET, the charge stability regions of the SET, the effect of nearby 'offset' charges, and energy band pictures are discussed. Finally, the chapter provides a brief introduction to quantum dots, where quantum confinement of electrons on the island leads to discrete energy levels on the island. These effects are of greater significance in devices where the island is defined using semiconductors. Our discussion considers the energy band diagram for a quantum dot, the 'addition energy' of the quantum dot, and the effect of quantum confinement effects on the electrical characteristics of the quantum dot.

Chapter 3, 'Single-Electron Transistors in Silicon', describes the design, fabrication and characterization of SETs in crystalline and nanocrystalline silicon material. The chapter begins with an introduction to early observations of single-electron effects in silicon, and the first silicon SET designs. The chapter then discusses SETs fabricated in crystalline silicon. SETs with lithographically defined islands are reviewed, including SETs fabricated in SOI material, with islands defined by etching and oxidization, room temperature SET designs and SETs based on MOSFET structures. Techniques such as pattern-dependant oxidation, used to define well-controlled islands ~10 nm in scale, are discussed. We then discuss SETs using silicon nanowires defined in SOI material. Here, fluctuations in the doping concentration, surface potential, etc. lead to the formation of MTJs along the nanowires. SETs with ultra-small islands ~5 nm or less in size, where room-temperature single-electron oscillations with high peak-to-valley ratios are observed, are introduced. The chapter then discusses the design, fabrication and characterization of a nanowire SET in SOI material, defined using electron-beam lithography. The electrical characteristics of the device, including room temperature operation, are discussed in detail.

The second part of the chapter discusses SETs in nanocrystalline silicon material, where the islands are defined 'naturally' by growth techniques rather than high-resolution lithography. The formation of potential barriers at the grain boundaries in nanocrystalline silicon thin films, and the conduction mechanism across such a film, are introduced first. This is then followed by a review of SETs in nanocrystalline silicon films, where the grains define islands and the grain boundaries define tunnel barriers. We discuss nanocrystalline silicon nanowire SETs, and scaling of these devices to form 'point-contact' SETs capable of room-temperature operation. We then discuss SETs where the islands are formed by discrete silicon nanocrystals. The observation of electrostatic and electron wavefunction coupling effects in nanocrystalline silicon SETs is considered. Finally, the chapter discusses single-electron effects in silicon nanowires and nanochains, synthesized by material growth processes rather than by lithographic techniques.

Chapter 4, 'Single-Electron Memory', discusses single-electron, and few-electron memory cells and circuits. In these devices, the Coulomb blockade effect is used to store information 'bits' consisting of single electrons or, at most, a few tens of electrons. The chapter begins with a brief historical introduction to charge storage in single-electron systems, and a discussion of the first single-electron memory design, the MTJ single-electron memory. Here, an MTJ is used to trap a small number of electrons on a memory node. The chapter discusses the concept of 'critical charge', and the hysteresis in the charge stored in the device. The chapter goes on to discuss MTJ memory designs in silicon, with various means to 'sense' the stored charge, e.g. using SETs or using scaled MOSFETs. A scaled MTJ memory cell, where one electron can be stored and sensed, is also discussed. This is followed by a brief review of single-electron memories using nanostructured 'floating gates' placed between insulating layers to store the charge. Here, the floating gate may be formed by a layer of silicon nanocrystals, or by a single, nanoscale floating gate. The silicon nanocrystals, or the scaled floating gate, may be small enough for room temperature single-electron effects. The chapter then discusses more complex memory designs, e.g. background charge insensitive single-electron memory, and a 128 Mb LSI single-electron memory in nanocrystalline silicon. Finally, the chapter discusses in detail the fabrication and characterization of an MTJ few-electron memory with MOSFET sensing, ~60 electrons per bit, and writing times ~10 ns.

Chapter 5, 'Few-Electron Transfer Devices', discusses the design, fabrication and operation of single-electron circuits capable of controlling the transfer of charge packets consisting of single electrons, or only a few electrons, using r.f. signals. The chapter begins with an introduction to the first single-electron transfer devices demonstrated, the single-electron turnstile and the single-electron pump. While these devices were implemented initially in the Al/AlO$_x$ and the GaAs/AlGaAs 2-DEG system and not in silicon, the design of other electron transfer systems relies on the concepts developed by these devices. The chapter then considers silicon MTJ-based bi-directional electron pumps and turnstiles, which may be regarded as MTJ analogues of the basic single-electron pump and turnstile. However, the charge packets in these devices consist typically of a few electrons and not one electron. Various designs of MTJ-based electron pumps/turnstiles are possible, using one or more r.f. signal. The chapter then considers single-electron transfer in other types of devices, e.g. nanoscale charge-coupled devices, and hybrid designs using integrated SETs and MOSFETs. Finally, the chapter briefly discusses metrological applications of single-electron transfer circuits.

Chapter 6, 'Single-Electron Logic Circuits', discusses systems where single, or at most a few electrons, are used to perform logic operations. Single-electron logic is possible using two approaches. One approach uses SETs as switching transistors in a manner similar to conventional logic circuits, and is referred to as 'voltage state' logic. An alternative approach uses a single electron, or a few-electron packet, to represent a bit. The physical presence of the electron packet at a given point in the logic circuit then represents a '1', and the absence of the packet represents a '0'. This approach is referred to as 'charge state' logic. The chapter first considers devices operating using voltage state logic. These include SET analogues of p and n metal oxide semiconductor (p-MOS and n-MOS) logic, and SET analogues of complementary metal oxide semiconductor (CMOS) logic. SET-based inverter, NAND and NOR logic gates, programmable SET-based logic gates, exclusive-OR (XOR) logic gates, and 'majority logic' gates are discussed. The chapter then considers devices operating using charge state logic. The most widely used scheme is binary decision diagram (BDD) logic, where charge packets are switched through a network formed by two-way switches into one of two output terminals. After an introduction to various BDD gates, the chapter discusses the design, fabrication and characterization of these devices in SOI material, using MTJ few-electron pumps. Finally, the chapter considers 'wireless' logic schemes such as the quantum cellular automaton (QCA) and the single-electron parametron, where arrays of cells formed by quantum dots, switched using electric fields, have been proposed for performing logic operations.

Chapter 2

Single-Electron Charging Effects

2.1 Introduction

Consider a system where a small conducting region, or 'island', is coupled by tunnel junctions to electrode regions. If the capacitance C_i of the island is small enough such that the charging energy of the island corresponding to even one electron, $E_c = e^2/2C_i$, has a significant value, then it is possible to precisely control the charging of the island at the one electron level. This process is referred to as the single-electron charging effect.

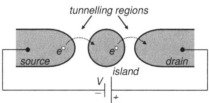

Fig. 2.1 Single-electron charging of an island isolated by tunnel barriers.

Figure 2.1 shows a schematic diagram of the island, connected by two tunnel junctions to source and drain electrodes. The island and the electrodes are conducting, and may be formed by either semiconductor or metal regions. In contrast, the tunnel junctions are relatively insulating. If a voltage is applied across such a system, electrons can tunnel across the first junction onto the island, and then off the island across the second junction, leading to current flow from the source to the drain. We assume that this process is fast enough such that the electrons charge the island one at a time, and the extra charge on the island is a discrete multiple of the single electron charge e. However, in order for even one electron to tunnel onto the island, it is necessary for the supply voltage to overcome the 'single-electron charging energy', $E_c = e^2/2C_i$. In a μm-scale or larger electronic device, C_i would be relatively large and E_c would have an insignificantly small value. However, in nanoscale devices, C_i can be small enough such that E_c is much more significant. It is then necessary for the applied voltage to be large enough to overcome E_c, before even one electron can tunnel onto the island and a current can flow. This leads to a 'Coulomb blockade' of conduction at low values of applied voltage. As each extra electron charges the island, the corresponding charging energy must be overcome, strongly influencing current flow across the island.

Single-electron effects have been identified since the 1950s, in measurements of the conduction and charging process in granular metal films, where the grain sizes were in the nanometre scale (Gorter, 1951; Neugebauer and Webb, 1962; Giaver and Zeller, 1968; Lambe and Jaklavic, 1969; Zeller and Giaver, 1969; Kulik and Shekter, 1975). In these films, at cryogenic temperatures, the capacitance of each grain could be small enough such that E_c was large compared to the thermal energy k_BT, leading to a Coulomb blockade of the conduction at low bias. Clearly, E_c must be greater than k_BT, otherwise thermal fluctuations in energy will overcome the single-electron charging energy, smearing out any single-electron effects. This gives the first condition for the observation of single-electron effects:

$$E_c \gg k_BT \qquad (2.1)$$

The first lithographically fabricated single-electron device as opposed to a 'naturally' formed granular system, the single-electron transistor (SET) of Fulton and Dolan (Fulton and Dolan, 1987) had island capacitances $C_i \sim 1$ fF, such that $E_c \sim 100$ μV. This corresponded to $T <$ 1 K, requiring that the measurements were performed in a dilution refrigerator at milli-Kelvin temperatures. Since the 1990s, with advances in nanolithography, it has now become possible to fabricate islands with capacitances $C_i \sim 10^{-18}$ F, raising $E_c \sim 0.1$ V and T to room temperature (Takahashi et al., 1995).

We also require that electrons can be localized on the island, such that an extra electron on the island can affect the tunnelling of further electrons onto the island. If the electron wavefunction is *delocalized* from one electrode to the other across the island, then a current can flow through such a delocalized state without increasing the island charge (Likharev and Zorin, 1985; Averin and Likarev, 1986; Zwerger and Scharpf, 1991). The requirement that electrons are localized on the island implies that the tunnel resistance R_T is relatively large, leading to the second condition for the observation of single-electron effects:

$$R_T \gg R_K = h/e^2 = 25.9 \text{ k}\Omega \qquad (2.2)$$

Here, R_K is the quantum of resistance. As a consequence of this condition, single-electron devices have relatively high resistance, at least compared to conventional CMOS devices. Equation (2.2) also implies that coherent 'co-tunnelling' processes (Averin and Odintsov, 1989; Geerligs *et al.*, 1990; Averin and Nazarov, 1992), consisting of several simultaneous tunnelling events, may be neglected. Finally, we assume that the time for an electron to tunnel through the tunnel barrier is small compared to any other time scale, e.g. the time between successive tunnelling events, implying that electrons tunnel one-by-one. This assumption, Eq. 2.1, and Eq. 2.2, form the basic requirements for single-electron charging, in the so-called 'orthodox theory' of single-electron effects (for reviews, see Averin and Likharev, 1991; Likharev, 1991; Ingold and Nazarov, 1992).

In most of the following discussion, we will consider the island, and the electrodes, to be metallic. As we will be concerned with semiconductor single-electron devices in following chapters, this assumption is valid only if electrons or holes are present in the conduction or valance band at the operating temperature of the device, i.e. partially filled bands are present. Many silicon single-electron devices are heavily-doped *n*- or *p*-type and, in this case, carriers exist in the conduction or valance band even at low temperatures (see Ali and Ahmed, 1994; Smith and Ahmed, 1997a). The device may then be regarded as metallic. In other devices, while the electrodes may be metallic, the island may be intrinsic, without any carriers at the operating temperature (see Takahashi *et al.*, 1995). Here, the potential of the island must be adjusted, often using a gate electrode, to allow conduction or valance band states to overlap with the Fermi energy in the electrodes. In this case, one may add electrons to the island using n-type electrodes, or add holes to the island using p-type electrodes (Ishikuro and Hiramoto, 1999). Furthermore, the island size may be small enough such that the quantization of electronic states on the island may be observed (see Ishikuro and Hiramoto, 1997). In such an island, usually referred to as a 'quantum dot' (Reed, 1991), the conduction mechanism depends not only on single-electron charging effects, but also on resonant tunnelling through the quantized states. Quantum dots will be considered briefly in Section 2.5.

There are a number of reviews of theoretical work on single-electron devices. For example, the subject was reviewed at an early stage by Likharev (Likharev, 1988), Averin and Likharev (Averin and Likharev, 1991), and Schön and Zaiken (Schön and Zaiken, 1990). The reader is also referred to the book *Single Charge Tunneling*, edited by Grabert and Devoret (Grabert and Devoret, 1992), especially the Introduction by Devoret and Grabert, and the paper by Ingold and Nazarov (Ingold and Nazarov, 1992). There is a more recent general review by Likharev (Likharev, 1999). These works and the references therein may be referred to for details beyond the scope of the present chapter.

2.2 Single Tunnel Junction

The simplest system where the charging energy associated with a single electron may affect the tunnelling of another electron is a single tunnel junction. In early work on single tunnel junctions, several new effects were predicted (Ben-Jacon and Gefen, 1985; Likharev and Zorin, 1985; Averin and Likarev, 1986). The most well-known of these effects is the single-electron oscillation of current at a fundamental frequency $f_{SET} = I/e$. The single tunnel junction is of considerable interest in investigating the physics of electron tunnelling, and also forms the basic element in more complicated systems such as the SET. Single-electron effects in this system can, however, be difficult to observe, mainly due to the effect of quantum fluctuations in the energy of the circuit environment. In contrast to this, in systems where an island is isolated by tunnel junctions or capacitors, the island is isolated from fluctuations in the environment and the observation of single-electron effects is straightforward. In the following, we consider the single tunnel junction only briefly. Detailed reviews of the system are given by Devoret and Grabert (Devoret and Grabert, 1992) and Ingold and Nazarov (Ingold and Nazarov, 1992).

Consider a single tunnel junction, with a capacitance C and a resistance R, biased by an ideal current source I (Fig. 2.2[a]). The voltage V across the junction is measured by a high-impedance voltmeter. The charge on the junction Q may be regarded as a continuous variable as it depends on the displacement of the electron sea in the electrodes, with respect to the positive ionic background. Q may have any value, including arbitrarily small fractions of the elementary charge e.

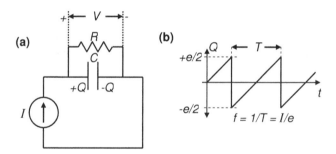

Fig. 2.2 The single tunnel junction. (a) Equivalent circuit. (b) Time-dependent single-electron tunnelling oscillations.

We now assume that electrons tunnel discretely across the tunnel junction, such that n electrons have tunnelled at a time t. We further assume that both n and Q are well-defined variables. The current I is then given by charge conservation, $I = Q'(t) + en'(t)$. When an electron tunnels across the junction, Q changes by e and the electrostatic energy E of the junction changes by:

$$\Delta E = E_{final} - E_{initial} = \frac{(Q-e)^2}{2C} - \frac{Q^2}{2C} = \frac{e}{C}\left(\frac{e}{2} - Q\right) \quad (2.3)$$

At temperature $T = 0$ K, ΔE must be negative for an electron to tunnel across the junction. An electron then tunnels across the junction if Q exceeds $e/2$. Assuming that initially, zero electrons have tunnelled across the junction, applying a constant current bias I leads to Q increasing linearly with time, as $dQ=Idt$. When Q exceeds the critical charge $e/2$, ΔE becomes negative and an electron tunnels across the junction. This causes Q to reduce by e, to $-e/2$, and the process begins again. The net result is sawtooth oscillations in Q, and in the junction voltage $V = Q/C$ (Fig. 2.2[b]), at a frequency:

$$f_{SET} = I/e \quad (2.4)$$

These oscillations, usually referred to as single-electron tunnelling oscillations, provide a fundamental relationship between frequency and current. Furthermore, the I-V characteristics show a zero current region in the range:

$$-\frac{e}{2C} < V < \frac{e}{2C} \quad (2.5)$$

Here, V depends on the charge on the junction, determined by the history of the current in the leads. This is the Coulomb blockade region for a single tunnel junction.

Single-electron effects in a single tunnel junction can often be difficult to observe experimentally, due to the effect of quantum fluctuation of charge in the circuit environment of the tunnel junction. A detailed quantum mechanical analysis of a tunnel junction and its environment will not be provided here and may be found in specific articles (Nazarov, 1989; Devoret *et al.*, 1990; Girvin *et al.*, 1990), and various reviews (Devoret and Grabert, 1992; Ingold and Nazarov, 1992). These works show that if the environmental impedance $Z(\omega) \ll R_K = 25.9$ kΩ, then charge fluctuations in the environment are much larger than e, and it is not possible to observe the charge transfer of single electrons. However, if $Z(\omega) \gg R_K$, then it is possible to observe single-electron effects at oscillation frequencies $f \leq \tau^{-1}$, where $\tau \sim h/(e^2/C) = R_K C$ is the uncertainty time associated with the Coulomb charging energy. For this condition, the tunnelling electron exchanges energy with the modes of the environment at an energy equal to the charging energy. This results in tunnelling of the electron once the charging energy is overcome. Unfortunately, practical realization of the condition can be difficult, as the resistance of the leads connecting to the tunnel junction must be high compared to R_K, but not so high that there is excessive heating. In an experimental realization of the system, Cleland *et al.* (Cleland *et al.*, 1990) observed single-electron effects in a single tunnel junction, using NiCr leads with resistances of 29 kΩ.

2.3 The Single-Electron Box

We have seen in the previous section that it is difficult to experimentally observe the Coulomb blockade of conduction through a single tunnel junction. This is because of the coupling of the tunnel junction to the environmental modes of the circuit, which overcome the Coulomb blockade if the environmental impedance is low. Single-electron tunnelling oscillations through the single tunnel junction are a dynamic effect, with successive tunnelling processes. Observation of the Coulomb blockade of a single tunnel junction requires the slowing down of the tunnel rates as much as possible, by increasing the environmental impedance (Devoret and Grabert, 1992).

The simplest circuit where Coulomb blockade effects can be observed at equilibrium is the single-electron box (Fig. 2.3) (Lafarge *et al.*, 1991; Devoret and Grabert, 1992). This consists of a tunnel junction with a capacitance C_1 and resistance R_1, in series with a storage capacitor C_b, isolating an island in-between. A voltage V may be used to bias the circuit such that electrons are transferred across the tunnel junction, on to or off C_b.

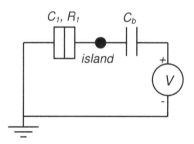

Fig. 2.3 The single electron box.

The presence of the additional capacitance C_b allows one to observe Coulomb blockade at equilibrium. Furthermore, C_b decouples the island charge ne from the environment. This is because the two capacitors couple as an equivalent capacitance $C_{eq} = C_1 C_b / (C_1 + C_b)$, with a charge $Q_{eq} = C_1 C_b V / (C_1 + C_b) = (Q_1 C_b + Q_b C_1)/(C_1 + C_b)$. Here, Q_1 and Q_b are the charges on C_1 and C_b, respectively. Only Q_{eq} is affected by the environment. The island charge is given by $ne = Q_1 - Q_b$, and this charge decouples from the leads (Grabert et al., 1991). This allows the experimental observation of Coulomb blockade in a low impedance environment as well. For the remainder of this chapter, we will consider a low impedance environment as standard.

2.3.1 The 'critical charge'

We begin by developing a simple picture for a single tunnel junction, embedded in a circuit (Fig. 2.4). The tunnel junction may be represented by a resistance R_1 corresponding to the tunnel resistance, in parallel with a capacitance C_1 corresponding to the junction capacitance (Fig. 2.4[a]). Here, R_1 allows electrons to be transferred across the tunnel junction, and C_1 allows a charge Q_1 to build up on the junction. We recall that Q_1 is a continuous variable and can be fractionally small, as it depends on the displacement of electrons in the electrodes relative to their ionic background. In order to determine the charging of the tunnel junction embedded in a circuit, we represent the circuit by an equivalent capacitance C_{ext}, in parallel with the tunnel element. This leads to the circuit of Fig. 2.4(b), where we omit R_1 for simplicity.

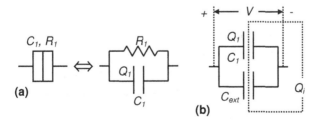

Fig. 2.4 The single tunnel junction. (a) Circuit model. (b) The tunnel capacitance C_1, in parallel with the external circuit capacitance C_{ext}.

The charging energy for the circuit of Fig. 2.4(b), for an equivalent charge Q_i on the parallel combination of C_1 and C_{ext}, is given by:

$$E_i = \frac{Q_i^2}{2(C_1 + C_{ext})} \qquad (2.6)$$

We now consider an electron entering the circuit, such that Q_i changes to $Q_i - e$. This leads to a change in the charging energy of the circuit, ΔE_i, given by:

$$\Delta E_i = E_{final} - E_{initial} = \frac{(Q_i - e)^2}{2(C_1 + C_{ext})} - \frac{Q_i^2}{2(C_1 + C_{ext})} \qquad (2.7)$$

With a voltage V_1 across the circuit, the charge on the tunnel junction C_1 can be related to the equivalent charge Q_i by:

$$V_1 = \frac{Q_1}{C_1} = \frac{Q_i}{C_1 + C_{ext}} \qquad (2.8)$$

Using Eq. 2.8 to eliminate Q_i from Eq. 2.7 gives us ΔE_i in terms of the tunnel junction charge Q:

$$\Delta E_i = \frac{e}{C_1}\left(\frac{eC_1}{2(C_1 + C_{ext})} - Q_1\right) = \frac{e}{C_1}(Q_c - Q_1) \qquad (2.9)$$

Here, the *critical charge* Q_c (Geerligs *et al.*, 1990; Grabert *et al.*, 1991; Nakazato *et al.*, 1994) is given by:

$$Q_c = \frac{e}{2(1 + C_{ext}/C_1)} \qquad (2.10)$$

For non-zero values of C_{ext}, $Q_c < e/2$. In a similar manner, for an electron leaving the circuit, Q_i changes to $Q_i + e$ and the change in electrostatic energy is given by $\Delta E_i = (e/C_1)(Q_c + Q_1)$. For an electron to tunnel across the tunnel junction C_1, ΔE_i must be negative for the process to be energetically favourable. If ΔE_i is positive, electron tunnelling is suppressed and Coulomb blockade of conduction occurs. This situation corresponds to a range:

$$-Q_c < Q_1 < Q_c \qquad (2.11)$$

Equations 2.10 and 2.11 provide a means to determine whether a tunnel junction is in Coulomb blockade or not. Here, one determines first the equivalent capacitance of the rest of the circuit, C_{ext}, and then the critical charge Q_c. This gives the range of charge Q_1, or junction voltage $V_1 = Q_1/C_1$, where tunnelling is suppressed and the circuit can be in Coulomb blockade. This corresponds to a voltage range:

$$\frac{-e}{2C_1(1+C_{ext}/C_1)} < V_1 < \frac{e}{2C_1(1+C_{ext}/C_1)} \quad (2.12)$$

We note that if $C_{ext} \sim 0$, then Eq. 2.11 reduces to $Q_c = e/2C$, and Eq. 2.12 reduces to $-e/2C_1 < V_1 < e/2C_1$, the results for an isolated single tunnel junction.

We now use this approach to calculate the critical charge for the single-electron box. For the simple circuit of Fig. 2.3, the island voltage is given by the tunnel junction voltage V_1. In addition, looking from the island C_b is in parallel with C_1, a configuration similar to Fig. 2.4. Equation 2.10 then gives the critical charge for an electron added to the island of the single-electron box, with C_{ext} replaced by C_b:

$$Q_c = \frac{e}{2(1+C_b/C_1)} = \frac{eC_1}{2(C_1+C_b)} \quad (2.13)$$

If the bias voltage V is increased, the island charge Q_1 increases until it exceeds the above value of Q_c and the Coulomb blockade is overcome. An electron then tunnels onto the island, changing Q_1 to $Q_1 - e$, which is less than Q_c. The Coulomb blockade is then re-imposed and the electron is trapped on the island. Further increases in V overcome the Coulomb blockade periodically and allow additional electrons to be trapped on the island.

2.3.2 Electrostatic energy changes

We will now calculate the change in the electrostatic energy of the single-electron box when an electron is added onto the island. We use an approach (Ingold and Nazarov, 1992) where we calculate the change in the charge of the capacitors C_1 and C_b when an electron tunnels onto the island. This creates a non-equilibrium condition, and equilibrium is reestablished by charge transfer from the supply voltage V. The total change in electrostatic energy is then given by the sum of the change in the island charging energy and the work done by V. This approach considers the change in the charging energy of the entire circuit, conforming to the so-called 'global' view of the system (Schön and Zaiken, 1990; Averin and Likharev, 1991; Geigenmüller and Schön, 1989; Likharev et al., 1989).

Figure 2.5(a) shows the circuit diagram of the single-electron box, biased by a supply voltage V. Looking from the island (Fig. 2.5[b]), C_1 and C_b are in parallel and the island charge $-ne$ is given by:

$$-ne = Q_b - Q_1 \tag{2.14}$$

Note that the term ne is positive and corresponds to n electrons on the island.

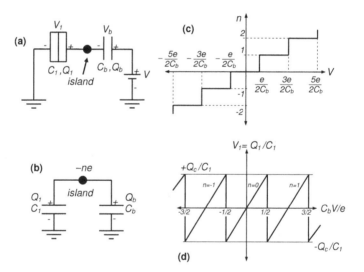

Fig. 2.5 Charge build-up in the single-electron box. (a) Circuit diagram. (b) Capacitances C_1 and C_b, looking from the island. (c) Island charge n vs. supply voltage V. (d) Tunnel junction voltage V_1 vs. supply voltage V.

Applying Kirchhoff's voltage law to the circuit:

$$V_1 + V_b - V = 0 \tag{2.15}$$

$$\Rightarrow \frac{Q_1}{C_1} + \frac{Q_b}{C_b} - V = 0 \qquad (2.16)$$

Solving Equations 2.14 and 2.16 simultaneously, we can find Q_1 and Q_2 in terms of *ne* and V:

$$Q_1 = \frac{C_1}{C_\Sigma}(C_b V + ne) \qquad (2.17a)$$

$$Q_b = \frac{C_b}{C_\Sigma}(C_1 V - ne) \qquad (2.17b)$$

where $C_\Sigma = C_1 + C_b$.

Now, consider an electron tunnelling onto the island, across the tunnel junction C_1, from the left to the right. Then the island charge changes from $-ne$ to $-ne-e$, i.e. n increases to $n+1$. This changes the tunnel junction charge Q_1 as follows:

$$\Delta Q_1 = Q_{1,final} - Q_{1,initial} = \frac{C_1}{C_\Sigma}(C_b V + (n+1)e) - \frac{C_1}{C_\Sigma}(C_b V + ne)$$

$$\Rightarrow \Delta Q_1 = \frac{C_1}{C_\Sigma} e \qquad (2.18a)$$

In a similar manner, the change in the storage capacitor charge Q_b is given by:

$$\Delta Q_b = Q_{b,final} - Q_{b,initial} = -\frac{C_b}{C_\Sigma} e \qquad (2.18b)$$

Note that Equations 2.18a and b imply that $\Delta Q_b - \Delta Q_1 = -e$, i.e. the island charge changes by $-e$, consistent with our argument.

We are now in a position to calculate the change in electrostatic energy of the circuit, ΔE_{add}, when an electron tunnels onto the island. This is given by the sum of the change in the electrostatic energy of the island, and the work done by the source V on the circuit:

$$\Delta E_{add} = E_{final} - E_{initial} + work = \frac{(-(n+1)e)^2}{2C_\Sigma} - \frac{(-ne)^2}{2C_\Sigma} + V\Delta Q_b$$

$$\Rightarrow \Delta E_{add} = \frac{e}{C_\Sigma}\left(-C_b V + ne + \frac{e}{2}\right) \qquad (2.19a)$$

where we have substituted the expression for ΔQ_b given by Eq. 2.18b.

In a similar manner, it is possible to calculate the change in electrostatic energy of the circuit, ΔE_{sub}, when an electron tunnels off the island, and the island charge changes from ne to $ne + e$. Note that for this transition, $\Delta Q_b = C_b e/C_\Sigma$. We then have the following expression for ΔE_{sub}:

$$\Delta E_{sub} = \frac{e}{C_\Sigma}\left(C_b V - ne + \frac{e}{2}\right) \qquad (2.19b)$$

Equations 2.19a and b can be used to calculate the tunnelling rate for electrons tunnelling on to or off the island. For a low impedance environment, the tunnelling rate for electrons tunnelling onto the island is given by (Devoret and Grabert, 1992):

$$\Gamma = \frac{1}{e^2 R_1} \frac{-\Delta E_{add}}{1 - \exp(\Delta E_{add}/k_B T)} \qquad (2.20a)$$

This formula is one of the main results of the 'orthodox theory' for single-electron effects (see Likharev, 1999). At $T = 0$ K, Eq. 2.20a reduces to:

$$\Gamma = \frac{-\Delta E_{add}}{e^2 R_1} \quad \text{if } \Delta E_{add} < 0 \qquad (2.20b)$$

$$\Gamma = 0 \quad \text{if } \Delta E_{add} > 0 \qquad (2.20c)$$

Using ΔE_{sub}, similar expressions can be written for the tunnelling rate for removing electrons from the island. These expressions are the first where the tunnel resistance R_1 appears. It is clear that R_1 is significant in determining the tunnelling rate for a transition in the number of electrons n, and not in determining the values V where transitions occur.

Equations 2.20a and b define the edge of the Coulomb blockade region at $T = 0$ K. Within the Coulomb blockade, an electron cannot tunnel onto the island as $\Delta E_{add} > 0$, i.e. the final charging energy plus the work done is greater than the initial charging energy. From Eq. 2.19a, this occurs when $C_b V < (n + \frac{1}{2})e$. Using a similar argument, an electron cannot tunnel off the island if $\Delta E_{sub} > 0$. From Eq. 2.19b, this occurs when $C_b V > (n - \frac{1}{2})e$. These expressions lead to the following range for Coulomb blockade, for $T = 0$ K:

$$(n - \frac{1}{2})e < C_b V < (n + \frac{1}{2})e \qquad (2.21)$$

Within this range, the number of electrons n trapped on the island is stable. For a given n, if V is changed such that it moves just outside the range predicted by Eq. 2.21, then n adjusts such that Eq. 2.21 is again satisfied. This leads to Fig. 2.5(c), where stable values of n can be seen in a plot of n vs. V.

We can compare ΔE_{add} in Eq. 2.19a to ΔE_i in the critical charge picture, given by Eq. 2.9. Equation 2.19a may be re-arranged into the following form:

$$\Delta E_{add} = \frac{e}{C_1}\left(\frac{eC_1}{2C_\Sigma} - \frac{C_1(C_b V - ne)}{C_\Sigma}\right) = \frac{e}{C_1}(Q_c - Q_1) \qquad (2.22a)$$

where we have used the expression for the critical charge Q_c given in Eq. 2.13 and

$$Q_1 = \frac{C_1(C_b V - ne)}{C_\Sigma} \qquad (2.22b)$$

It can be seen that increasing V increases the tunnel junction charge Q_1 until Q_c is reached and an electron tunnels onto the island. Figure 2.5(d) shows the island voltage $V_1 = Q_1/C_1$, as a function of supply voltage V, using Eq. 2.22b. The island voltage varies between the limits $\pm Q_c/C_1$, e.g. if V increases, V_1 increases until the limit $+Q_c/C_1$ is reached. An electron then tunnels onto the island, increasing n to $n+1$ and reducing V_1 to $-Q_c/C_1$. Increasing V adds electrons to the island, and reducing V removes electrons from the island. We note, however, that there is no hysteresis in V_1. While the single-electron box may be regarded as a device that can hold a precise number of electrons, it is necessary to apply a voltage to keep the electrons in the box, e.g. in order to hold one extra electron, the applied voltage must lie in the range $e/2C_b < V < 3e/2C_b$. In order to form a *memory* device, charge must be retained even if the voltage is removed, requiring a hysteresis in the value of the charge as a function of the applied voltage. However, the concepts developed in the single-electron box can be applied to form single-electron traps and single-electron memory cells (Fulton et al., 1991; Nakazato et al., 1994). In these devices, the single tunnel junction of the single-electron box is replaced with a double, or a multiple-tunnel junction. We will discuss single-electron memory in detail in Chapter 4.

2.3.3 Energy diagram for the single-electron box

As the supply voltage V across the single-electron box is varied, electrons are added or removed from the island, with the charge state of the island expressed by an electron number n. Each value of n corresponds to a single-electron energy E_n, and if this energy is below the Fermi energy of the left-hand side 'ground' or source electrode, then n charging electrons exist on the island. It is possible to sketch an energy diagram for the single-electron box, showing the positions of E_n calculated using Equations 2.19a and b. Equation 2.19a gives the energy needed for an electron to be added to the island across the tunnel junction, with the initial island state n. For $V = 0$ and $n = 0$, a total energy $\Delta E_{add} = e^2/2C_\Sigma$ must be provided for an electron to be added to the island. This gives the energy for the first electron to be added to the island, $E_1 = e^2/2C_\Sigma$. Now, adding a second electron to the island is equivalent to adding an electron to the island in a state $n = 1$. For $n = 1$, Eq. 2.19a then gives $E_2 = 3e^2/2C_\Sigma$. Similarly, $n = 2, 3\ldots$ give the energies E_3, E_4, etc. Equation 2.14b may be used to give the energies E_{-1}, E_{-2}, E_{-3}, etc. This allows us to draw an energy diagram for the single-electron box, shown in Fig. 2.6(a). At $V = 0$, the states for $n > 0$ are empty, and the states for $n < 0$ are filled.

This energy diagram provides a simple picture to understand the charging of the island. If a voltage $+V$ is applied, then a bias drops across C_1 and C_b, lowering the positions of E_n relative to the Fermi energy E_{FS} in the source electrode (Fig. 2.6[b]). Here, voltages $V_{tj} = C_b V/C_\Sigma$ and $V_b = C_1 V/C_\Sigma$ drop across the tunnel junction and the storage capacitor, respectively. Therefore, at a voltage $V_{tj} = V_1 = e/2C_\Sigma$, corresponding to $V = e/2C_b$, E_1 is pulled level with E_{FS}. An electron then tunnels onto the island from the source, charging the island by one electron to the state

$n = 1$. As V is increased to $V_2 = 3e/2C_b$, E_2 is pulled level with E_{FS}, causing n to increase to 2. Further increases in V lead to increasing values of n, and all the states E_n lying below E_{FS} are filled. Similarly, a negative value of V reduces n, and the diagram corresponds to the plot of Fig. 2.5(d).

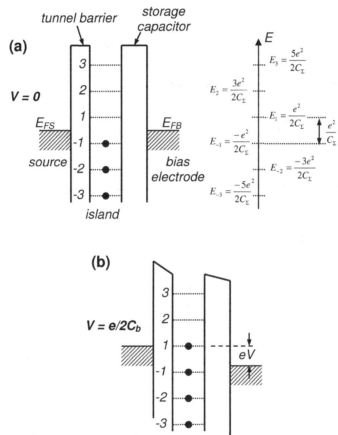

Fig. 2.6 Energy diagram for the single-electron box. (a) Zero applied bias $V = 0$. (b) With applied bias V.

2.4 The Single-Electron Transistor

We will now consider the single island SET (Fig. 2.7[a]) (Fulton and Dolan, 1987; Likharev, 1987; Likharev, 1988). The device consists of an island, isolated by two tunnel junctions from source and drain voltages. The left-hand side (Junction 1) and right-hand side (Junction 2) tunnel junction resistances and capacitances are R_1 and C_1, and R_2 and C_2, respectively. In order to simplify the analysis, we choose source and drain voltages $-V/2$ and $+V/2$ respectively, equal in magnitude and opposite in polarity. The voltages drive electrons from the left-hand side source terminal, across the two tunnel junctions, to the right-hand side drain terminal. The island is also coupled to an additional gate voltage V_g by a gate capacitor C_g. The addition of the gate capacitor converts the two-terminal double tunnel junction into an SET.

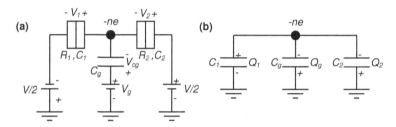

Fig. 2.7 The SET. (a) Circuit diagram. (b) Capacitances C_1, C_2 and C_g, looking from the island.

We analyse the device using a procedure (Ingold and Nazarov, 1992) similar to that used for the single-electron box. We calculate first the change in electrostatic energy of the entire circuit, for an electron tunnelling onto the island. Because of the two tunnel junctions here, the electron number on the island can change for an electron tunnelling on to or off Junction 1, or through a similar process across Junction 2. This leads to two equations which define the limits of the Coulomb blockade for n extra electrons on the island. Once the electrostatic energy changes are calculated, we use these to determine the backward and forward tunnel rates for each junction. Using the net tunnelling rate, and the probability of n extra electrons on the island, the device I-V characteristics can be predicted. As we shall see, the characteristics at T = 0 K show a zero current Coulomb blockade region, modulated in width by V_g. In addition, for mismatched tunnelling rates across the two junctions, the current rises in a step-like manner, creating a Coulomb staircase characteristic.

2.4.1 Electrostatic energy changes

We begin by calculating the change in electrostatic energy when an electron tunnels on to or off the island across the two tunnel junctions. Looking from the island (Fig. 2.7[b]), C_1, C_2 and C_g are in parallel, and the island charge $-ne$ is given by:

$$-ne = Q_2 + Q_g - Q_1 \qquad (2.23)$$

Note that, as before, the term ne is positive and corresponds to n electrons on the island.

Applying Kirchhoff's voltage law to the two loops formed in the circuit, we obtain:

Left-hand side loop:

$$-\frac{V}{2} + V_1 + V_{cg} - V_g = 0$$

$$\Rightarrow -\frac{V}{2} + \frac{Q_1}{C_1} + \frac{Q_g}{C_g} - V_g = 0 \qquad (2.24)$$

Right-hand side loop:

$$V_g - V_{cg} + V_2 - \frac{V}{2} = 0$$

$$\Rightarrow -\frac{V}{2} + \frac{Q_2}{C_2} - \frac{Q_g}{C_g} + V_g = 0 \qquad (2.25)$$

Solving Equations 2.23–2.25 simultaneously, we can find Q_1, Q_2 and Q_g in terms of ne, V_g and V:

$$Q_1 = \frac{C_1}{C_\Sigma}\left(\left(C_2 + \frac{C_g}{2}\right)V + C_g V_g + ne\right) \qquad (2.26a)$$

$$Q_2 = \frac{C_2}{C_\Sigma}\left(\left(C_1 + \frac{C_g}{2}\right)V - C_g V_g - ne\right) \qquad (2.26b)$$

$$Q_g = \frac{C_g}{C_\Sigma}\left((C_1 - C_2)\frac{V}{2} + (C_1 + C_2)V_g - ne\right) \qquad (2.26c)$$

where $C_\Sigma = C_1 + C_2 + C_g$.

Now, consider an electron tunnelling onto the island, across the tunnel junction C_1, from the left to the right. Then the island charge changes from $-ne$ to $-ne - e$, i.e. n increases to $n + 1$. This changes the tunnel junction charge Q_1 as follows:

$$\Delta Q_1 = Q_{1,final} - Q_{1,initial} =$$

$$\frac{C_1}{C_\Sigma}\left(\left(C_2 + \frac{C_g}{2}\right)V + C_g V_g + (n+1)e\right)$$

$$-\frac{C_1}{C_\Sigma}\left(\left(C_2 + \frac{C_g}{2}\right)V + C_g V_g + ne\right)$$

$$\Rightarrow \Delta Q_1 = \frac{C_1}{C_\Sigma}e \qquad (2.27\text{a})$$

In a similar manner, the changes in charges Q_1 and Q_g are given by:

$$\Delta Q_2 = -\frac{C_2}{C_\Sigma}e \qquad (2.27\text{b})$$

$$\Delta Q_g = -\frac{C_g}{C_\Sigma}e \qquad (2.27\text{c})$$

Note that Equations 2.27a, b and c imply that $\Delta Q_2 + \Delta Q_g - \Delta Q_1 = -e$, i.e. the island charge changes by $-e$, consistent with our argument.

We are now in a position to calculate the change in electrostatic energy of the circuit, $\Delta E_{1,add}$, when an electron tunnels onto the island across Junction 1 and n changes to $n + 1$. This is given by the sum of the change in the electrostatic energy of the island, and the work done by the left-hand side, right-hand side and gate voltages. The work done by each voltage source is given by the magnitude of the source, multiplied by the change in the charge of the associated capacitor. While the charges on C_2 and C_g change by ΔQ_2 and ΔQ_g, the charge on C_1 changes by $\Delta Q_2 - e$, as the total change in charge consists of the change in the electrostatic charge balance ΔQ_2, plus the transfer of an electron across C_1. Using Equations 2.27a, b and c, $\Delta E_{1,add}$ is then given by:

$$\Delta E_{1,add} = E_{final} - E_{initial} + work =$$

$$\frac{(-(n+1)e)^2}{2C_\Sigma} - \frac{(-ne)^2}{2C_\Sigma}$$

$$+ \frac{V}{2}(\Delta Q_1 - e) + \frac{V}{2}\Delta Q_2 + V_g \Delta Q_g$$

$$\Rightarrow \Delta E_{1,add} = \frac{e}{C_\Sigma}\left(ne + \frac{e}{2} - V(C_2 + \frac{C_g}{2}) - C_g V_g\right) \quad (2.28a)$$

In a similar manner, it is possible to calculate the change in electrostatic energy of the circuit, $\Delta E_{1,sub}$, when an electron tunnels off the island across Junction 1, and n changes to $n - 1$. We then have the following expression for $\Delta E_{1,sub}$:

$$\Delta E_{1,sub} = \frac{e}{C_\Sigma}\left(-ne + \frac{e}{2} + V(C_2 + \frac{C_g}{2}) + C_g V_g\right) \quad (2.28b)$$

In a manner analogous to the single-electron box, Equations 2.28a and b define the limits of the Coulomb blockade region for tunnelling across the first junction at $T = 0$ K, as a function of V and V_g and with n electrons on the island. If either $\Delta E_{1,add}$ or $\Delta E_{1,sub}$ are positive, then the final energy of the circuit, after the tunnelling of an electron on or off the island, is greater than the initial energy of the circuit and tunnelling is suppressed. Two edges of the Coulomb blockade region, for tunnelling across the first junction, can then be defined by the following inequality:

$$e(n - \frac{1}{2}) < V(C_2 + \frac{C_g}{2}) + C_g V_g < e(n + \frac{1}{2}) \quad (2.29a)$$

A similar calculation may be performed for an electron tunnelling across the second tunnel junction C_2, to obtain the corresponding energy changes $\Delta E_{2,add}$ and $\Delta E_{2,sub}$. For tunnelling across the second junction, two further edges of the Coulomb blockade region can be defined by the following inequality:

$$e(n - \frac{1}{2}) < -V(C_1 + \frac{C_g}{2}) + C_g V_g < e(n + \frac{1}{2}) \quad (2.29b)$$

Taken together, Equations 2.29a and b define the limits of the Coulomb blockade region for tunnelling across any of the two tunnel junctions. Within the range defined by these equations, the number of electron n on the island is stable.

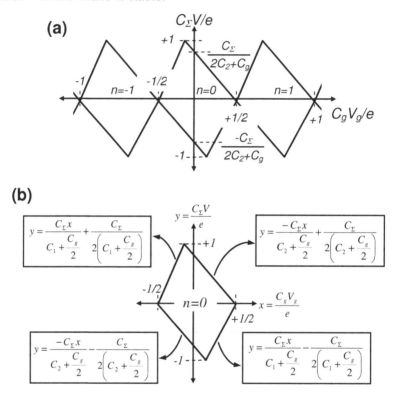

Fig. 2.8 Coulomb blockade in the SET. (a) Charge stability diagram, forming 'Coulomb diamonds'. (b) Edges of the $n = 0$ charge stability region.

Figure 2.8(a) plots the edges of the Coulomb blockade regions using Equations 2.29a and b and as a function of $C_g V_g/e$ and $C_\Sigma V/e$, for $n = -1$, 0 and +1. The Coulomb blockade regions are rhombic-shaped, each region corresponding to a specific, stable state of the electron number n. The regions are referred to as stability regions, or, because of their shape, as Coulomb 'diamonds'. The four inequalities in Equations 2.29a and b define the equations for the four edges of the stability region. Figure 2.8(b) shows these equations for the stability region for $n = 0$.

2.4.2 Tunnelling rates

We now consider the tunnelling rates across the two tunnel junctions, as a function of the energy changes $\Delta E_{1,add}$, $\Delta E_{1,sub}$, $\Delta E_{2,add}$ and $\Delta E_{2,sub}$. Initially, for simplicity we assume zero gate capacitance C_g. A detailed analysis, including the gate and any nearby trapped charge, is given by Ingold *et al.* (Ingold *et al.*, 1991). Here, we remove C_g, and the circuit of Fig. 2.7(a) reduces to that of Fig. 2.5(a). The energy changes (using Equations 2.28a and b and similar equations for Junction 2) for an electron tunnelling on to or off the island are then given by:

$$\Delta E_{1,add} = \frac{e}{C_\Sigma}\left(ne + \frac{e}{2} - C_2 V\right) \quad (2.30a)$$

$$\Delta E_{1,sub} = \frac{e}{C_\Sigma}\left(-ne + \frac{e}{2} + C_2 V\right) \quad (2.30b)$$

$$\Delta E_{2,add} = \frac{e}{C_\Sigma}\left(ne + \frac{e}{2} + C_1 V\right) \quad (2.30c)$$

$$\Delta E_{2,sub} = \frac{e}{C_\Sigma}\left(-ne + \frac{e}{2} - C_1 V\right) \quad (2.30d)$$

where $C_\Sigma = C_1 + C_2$.

Inspecting these equations, it is seen that $\Delta E_{1,sub}(V, ne) = \Delta E_{1,add}(-V, -ne)$ and $\Delta E_{2,sub}(V, ne) = \Delta E_{2,add}(-V, -ne)$. Simplifying our notation and replacing $\Delta E_{1,add}$ by ΔE_1, and $\Delta E_{2,add}$ by ΔE_2, we may replace $\Delta E_{1,sub} = \Delta E_1(-V, -ne)$ and $\Delta E_{2,sub} = \Delta E_2(-V, -ne)$.

We use Equations 2.30a, b, c and d to calculate the tunnelling rates for an electron tunnelling from left to right, and from right to left, across each of the tunnel junctions. For a low impedance environment, the tunnelling rate across the first junction, tunnelling from left to right, is given by:

$$\Gamma_{1,l-r} = \frac{1}{e^2 R_1} \frac{-\Delta E_1}{1 - \exp(\Delta E_1 / k_B T)} \quad (2.31)$$

At $T = 0$ K, this is non-zero when:

$$\Gamma_{1,l-r} = \frac{-\Delta E_1(V, ne)}{e^2 R_1} \quad \text{if } \Delta E_1(V, ne) < 0 \quad (2.32a)$$

Similarly, the tunnelling rate at $T = 0$ K for tunnelling from right to left, across the first junction is non-zero for $\Delta E_{1,sub} = \Delta E_1(-V, -ne) < 0$ and is given by the equation below. Note that this corresponds to an electron removed from the island.

$$\Gamma_{1,r-l} = \frac{-\Delta E_1(-V, -ne)}{e^2 R_1} \quad \text{if } \Delta E_1(-V, -ne) < 0 \quad (2.32b)$$

The two corresponding tunnelling rates across the second tunnel junction are given by:

$$\Gamma_{2,r-l} = \frac{-\Delta E_2(V, ne)}{e^2 R_2} \quad \text{if } \Delta E_2(V, ne) < 0 \quad (2.33a)$$

$$\Gamma_{2,l-r} = \frac{-\Delta E_2(-V, -ne)}{e^2 R_2} \quad \text{if } \Delta E_2(-V, -ne) < 0 \quad (2.33b)$$

Here, the Eq. 2.33a corresponds to an electron added to the island, and Eq. 2.33b corresponds to an electron removed from the island. The direction of each of the tunnelling rates in Equations 2.32a, b and c, and Equations 2.33a and b are marked in Fig. 2.9.

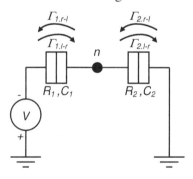

Fig. 2.9 Tunnelling rates across the double tunnel junction.

Now, using Equations 2.30a, b, c and d, 2.32a and b, and 2.33a and b, it may be seen that the energy changes in Equations 2.32 and 2.33 are less than zero, and the tunnelling rates are non-zero, for the following conditions:

$$\Gamma_{1,l-r} \neq 0: \quad +ne + \frac{e}{2} - C_2V < 0 \qquad (2.34a)$$

$$\Gamma_{1,r-l} \neq 0: \quad -ne + \frac{e}{2} + C_2V < 0 \qquad (2.34b)$$

$$\Gamma_{2,r-l} \neq 0: \quad +ne + \frac{e}{2} + C_1V < 0 \qquad (2.34c)$$

$$\Gamma_{2,l-r} \neq 0: \quad -ne + \frac{e}{2} - C_1V < 0 \qquad (2.34d)$$

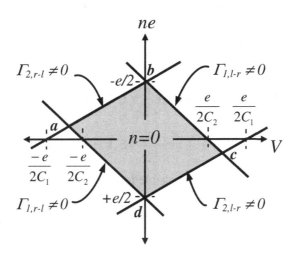

Fig. 2.10 Tunnelling rates in a double tunnel junction, at the boundary of the Coulomb blockade region.

Equations 2.34a, b, c and d are plotted in Fig. 2.10 as a function of V and ne, where we assume that $C_1 < C_2$. The equations define the edges of the Coulomb blockade (shaded region) with $n = 0$. This region is 'tilted' relative to the x-axis, i.e. the points a and c do not lie on the x-axis. This is a consequence of the different values of C_1 and C_2, and if $C_1 = C_2$, then a and b lie on the x-axis. Within the Coulomb blockade region, the state $n = 0$ is stable as all the tunnelling rates are zero. Outside this region, at least one of the rates becomes non-zero and an election tunnels on or off the island. For example, along the x-axis, $ne = 0$ and if V lies in the range $e/2C_2 < V < e/2C_1$, then only the tunnel rate $\Gamma_{1,l\text{-}r}$ is non-zero and an electron tunnels onto the island across Junction 1. Similarly, for $ne = 0$ and the range $-e/2C_2 < V < -e/2C_1$, only the tunnel rate $\Gamma_{1,r\text{-}l}$ is non-zero and an electron tunnels off the island across Junction 1. Without assuming $C_1 < C_2$, the rates become non-zero for $|V| > \min(e/2C_1, e/2C_2)$. Finally, if $n \neq 0$, and $|V| < \min(e/2C_1, e/2C_2)$, then the tunnelling rates become non-zero in a way such that electrons are added or removed from the island until $n = 0$ and there are no extra electrons on the island.

We now extend the analysis to the SET (Fig. 2.7[a]). For an SET where $C_g \ll C_1$ and C_2, the tunnelling rates are non-zero when:

$$\Gamma_{1,l\text{-}r} \neq 0: \quad +ne + \frac{e}{2} - C_2 V - C_g V_g < 0 \qquad (2.35a)$$

$$\Gamma_{1,r\text{-}l} \neq 0: \quad -ne + \frac{e}{2} + C_2 V + C_g V_g < 0 \qquad (2.35b)$$

$$\Gamma_{2,r\text{-}l} \neq 0: \quad +ne + \frac{e}{2} + C_1 V - C_g V_g < 0 \qquad (2.35c)$$

$$\Gamma_{2,l\text{-}r} \neq 0: \quad -ne + \frac{e}{2} - C_1 V + C_g V_g < 0 \qquad (2.35d)$$

Using these equations, it is possible to plot the stability regions of the SET. Figure 2.11(a) plots the stability region $n = 0$ as a function of V and $C_g V_g$, for $C_1 < C_2$. The stability region is similar in shape to the corresponding plot for the simple double tunnel junction shown in Fig. 2.10. Again, the stability region is 'tilted', as $C_1 \neq C_2$. In Equations 2.35a, b, c and d, the additional term $C_g V_g$ acts analogous to an additional charge $q = C_g V_g$ on the island. The effect of this is to reduce the width of the stability region, i.e. the width of the Coulomb blockade gap in V, such that as $|C_g V_g|$ increases from zero to $e/2$, the Coulomb gap reduces from e/C_2 to zero.

Figure 2.11(b) shows a plot of Equations 2.35a, b, c and d for different n, forming the stability diagram of the SET. This consists of lines corresponding to the boundaries where the transition from a zero to a non-zero tunnelling rate is observed. The corresponding tunnel rates which become non-zero are marked on the figure. The lines intersect and form additional stability regions with different values of n, similar in shape to the stability region for $n = 0$ (Region A). Along the y-axis, the presence of the gate allows the observation of a series of stable states n, in Coulomb blockade at equilibrium. Within these stability regions, the tunnel current is zero. For example, within Region B, $C_g V_g$ lies in the range $e/2 < C_g V_g < 3e/2$, and all the tunnelling rates of Equations 2.35a, b, c and d are zero for $n = 1$. This implies that one electron has been added to the island by the gate voltage. Varying V_g at $V = 0$ allows transitions to stable states with different values of n. For $|C_g V_g| = (m+1/2)e$, where m is an integer, the Coulomb blockade width reduces to zero.

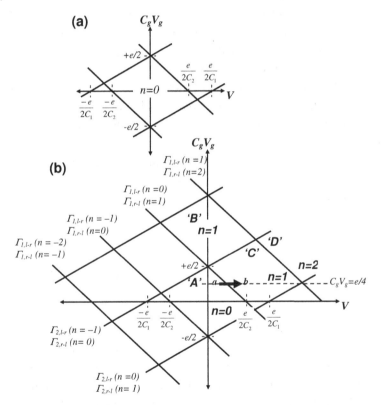

Fig. 2.11 Tunnelling rates in a SET, at the boundary of the Coulomb blockade region. (a) The $n = 0$ stability region, for $C_1 < C_2$. (b) Charge stability diagram for different n.

We will now consider the stability regions not along the y-axis, e.g. Region C where $n = 1$. If V or V_g are varied, such that the bias point of the SET moves from an adjacent stability region into Region C across the boundary between the two regions, then one of the tunnel rates becomes non-zero and n changes by 1. We now consider such a transition, e.g. from Region A to Region C along the line ab (Fig. 2.11[b]). Within Region A, $n = 0$ and all the tunnel rates are zero. Crossing the boundary between Regions A and C causes $\Gamma_{1,l\text{-}r} \neq 0$, while all the other rates remain zero. An electron then tunnels onto the island across tunnel Junction 1, changing n from 0 to 1. For $n = 1$, all rates except $\Gamma_{2,l\text{-}r}$ become zero. As $\Gamma_{2,l\text{-}r} \neq 0$, the electron then tunnels off the island, across tunnel Junction 2. This returns the SET to the $n = 0$ state, and the process can begin again. The net result is the sequential transfer of electrons from the left electrode to the right electrode.

As the two tunnelling processes occur sequentially, the total rate is dominated by the lower of the two rates. On average, the island charges up by one electron, and $n = 1$. It is not possible to charge the island with a second electron, unless the biases are increased such that a transition occurs to a stability region where $n = 2$. In a similar manner, for other stability regions not on the y-axis, the tunnelling rates add and remove electrons from the island in such a manner that the average number of extra electrons on the island is n.

2.4.3 Offset charge

We now consider the effect of the so-called 'offset' charge on the stability diagram of the SET (Ingold *et al.*, 1991). This charge consists of the charge induced on the island due to the trapped charges in the SET, near the island. As the offset charge is simply an induced charge, created by a shift in the electron distribution on the island, relative to the fixed ionic cores, it is not quantized and may have a value which is a fraction of e. An 'offset' charge q_0 simply changes the island charge from ne to $ne + q_0$. Replacing ne in Equations 2.35a, b, c and d by $ne + q_0$ allows us to define the corresponding ranges of non-zero tunnelling rates for the SET. The effect is to shift the stability diagram along the y-axis, negative values of q_0 shifting the stability diagram downwards, e.g. $q_0 = -e/2$

shifts the entire stability diagram of Fig. 2.11(b) downwards along the y-axis by $e/2$ and reduces the Coulomb blockade width near the origin to zero. As the offset charge can randomly vary from device to device, depending on the trapped charges, it may create problems in the reproducibility of the characteristics between nominally similar SETs. This complicates the practical implementation of these devices. The use of semiconductor SETs with passivated trap states, and the use of SETs with multiple-tunnel junctions, may reduce this problem. This is discussed in more detail in Chapter 3.

2.4.4 Calculation of I-V characteristics

The tunnelling rates of Equations 2.32a, b, c and d may be used to calculate the *I-V* characteristics at $T = 0$ of the double tunnel junction. For the SET, these tunnelling rates must use expressions for the changes in electrostatic energy ΔE_1 and ΔE_2 which also include the gate voltage V_g and gate capacitance C_g. These are straightforward to derive from Equations 2.29a and b. For $T > 0$, it is necessary to use the formula given in Eq. 2.31a to calculate the tunnelling rates.

As we have seen, the tunnelling rates depend on V, V_g and n. However, there is also a probability, p_n, that the island is in the state n, at each value of V and V_g. The current across the circuit may then be calculated by subtracting the backward tunnelling rate from the forward tunnelling rate, for a junction at a given value of V, V_g and n, multiplying the result by p_n and the elementary charge e, and summing this over all states n. This process gives the current at a given value of V and V_g, and can be repeated for a range of values of V or V_g to predict the *I-V* or the *I-V_g* characteristics. Here we assume that electrostatic equilibrium is restored in the circuit before each tunnelling event.

We begin by calculating p_n at constant bias V and V_g. Neglecting correlated tunnelling processes, p_n may change by leaving the state n, or by entering the state n, from the adjacent states $n + 1$ and $n - 1$. We can then write a master equation:

$$\frac{dp_n}{dt} = \Gamma_{n,n+1} p_{n+1} + \Gamma_{n,n-1} p_{n-1} - \Gamma_{n+1,n} p_n - \Gamma_{n-1,n} p_n \quad (2.36)$$

Here, $\Gamma_{l,m}$ is the rate of transition from state m to state l. The first two terms of the master equation then give the rate of change of the probability, from the state n to the adjacent states $n + 1$ and $n - 1$, and the last two terms give the rate of change of the probability from the adjacent states to state n. As state n may be changed by tunnelling across either of the two tunnel junctions, we may write $\Gamma_{l,m}$ in terms of the backwards and forwards tunnelling rates across the two tunnel junctions:

$$\Gamma_{n+1,n} = \Gamma_{1,l-r}(n) + \Gamma_{2,r-l}(n) \tag{2.37a}$$

$$\Gamma_{n-1,n} = \Gamma_{1,r-l}(n) + \Gamma_{2,l-r}(n) \tag{2.37b}$$

Here the backwards and forwards tunnelling rates are functions of n, and we have not explicitly shown the dependence on V and V_g, as they are constant. Now, as only adjacent states n and $n + 1$, or n and $n - 1$, are connected by non-zero tunnelling rates, a solution to the master equation is the following:

$$\Gamma_{n,n+1} p_{n+1} = \Gamma_{n+1,n} p_n \tag{2.38}$$

Here, the possibility of a transition from $n + 1$ to n equals the possibility of a transition from n to $n + 1$. Given a state $n = 0$, we may then calculate p_n as follows:

$$p_n = p_0 \prod_{m=0}^{m=n-1} \frac{\Gamma_{m+1,m}}{\Gamma_{m,m+1}} \tag{2.39a}$$

Similarly:

$$p_{-n} = p_0 \prod_{m=n+1}^{m=0} \frac{\Gamma_{m-1,m}}{\Gamma_{m,m-1}} \tag{2.39b}$$

The normalization of p_n gives a final equation:

$$\sum_{n=-\infty}^{n=+\infty} p_n = 1 \tag{2.39c}$$

Equations 2.39a, b and c may be solved simultaneously to find p_0 and then Equations 2.39a and b may be used to find all the remaining p_n. Finally, it is possible to calculate the current I at a given value of V and V_g:

$$I = e \sum_{n=-\infty}^{n=+\infty} p_n \left(\Gamma_{1,l-r}(n) - \Gamma_{1,r-l}(n) \right) =$$
$$e \sum_{n=-\infty}^{n=+\infty} p_n \left(\Gamma_{2,l-r}(n) - \Gamma_{2,r-l}(n) \right)$$

(2.40)

Repeating the process for a range of V and V_g can be used to calculate the I-V or the I-V_g characteristics. For low values of V and $T = 0$ K, only a small number of states are involved and the calculation of p_n is straightforward, e.g. for the stability diagram of Fig. 2.11(a), for $V_g = 0$ and V within the range $e/2C_2 < V < e/2C_1$, the transition from $n = 0$ to 1 across Junction 1 occurs first, followed by the transition from $n = 1$ to 0 across Junction 2. Only two states, $n = 0$ and $n = 1$ are then involved. Equation 2.37 gives:

$$\Gamma_{1,0} = \Gamma_{1,l-r}(0) + \Gamma_{2,r-l}(0) = \Gamma_{1,l-r}(0) \quad (2.41a)$$

$$\Gamma_{0,1} = \Gamma_{1,r-l}(1) + \Gamma_{2,l-r}(1) = \Gamma_{2,l-r}(1) \quad (2.41b)$$

Equation 2.39a then gives:

$$p_1 = p_0 \frac{\Gamma_{1,0}}{\Gamma_{0,1}} = p_0 \frac{\Gamma_{1,l-r}(0)}{\Gamma_{2,l-r}(1)}$$

Using this expression and Eq. 2.39c, we obtain:

$$p_0 + p_0 \frac{\Gamma_{1,l-r}(0)}{\Gamma_{2,l-r}(1)} = 1$$

$$\Rightarrow p_0 = \frac{\Gamma_{2,l-r}(1)}{\Gamma_{1,l-r}(0) + \Gamma_{2,l-r}(1)} \quad (2.42a)$$

$$\text{and } p_1 = \frac{\Gamma_{1,l-r}(0)}{\Gamma_{1,l-r}(0) + \Gamma_{2,l-r}(1)} \quad (2.42b)$$

The rates $\Gamma_{1,l-r}(0)$ and $\Gamma_{2,l-r}(1)$ depend on V. Using Eq. 2.40, the current is then given by:

$$I = e\Gamma(V) \quad (2.43a)$$

where the net rate $\Gamma(V)$ is given by:

$$\frac{1}{\Gamma(V)} = \frac{1}{\Gamma_{1,l-r}(0,V)} + \frac{1}{\Gamma_{2,l-r}(1,V)} \quad (2.43b)$$

Clearly, $\Gamma(V)$ is dominated by the lower of the rates $\Gamma_{1,l-r}(0, V)$ and $\Gamma_{2,l-r}(1, V)$. The junction with the lower rate acts as the bottleneck for the circuit. For higher values of V, additional transitions become possible and more tunnelling rates become non-zero. A numerical solution of Equations 2.39a, b and c is usually necessary to calculate the various probabilities p_n (see Amakawa et al., 1998) The I-V characteristics of an SET, and of more complicated systems, may also be calculated using a Monte Carlo approach, based on general formulas of the form of Eq. 2.20a (Likharev et al., 1989; Wasshuber et al., 1997; Amakawa et al., 1998).

2.4.5 The Coulomb staircase

The I-V characteristics of a simple double tunnel junction (Fig. 2.9), or of the SET (Fig. 2.7[a]) at a constant gate voltage V_g, show a Coulomb gap at low values of V. The Coulomb gap width in V corresponds to the width of the charge stability region at $n = 0$ for the double tunnel junction, or the width of the charge stability region at the applied value of V_g, for the SET. Outside the Coulomb blockade region, a non-zero current is observed, increasing in magnitude for increasing values of |V|. The manner in which this current increases depends on the tunnelling resistances of the two junctions, R_1 and R_2, respectively

If R_1 is very different from R_2, then the tunnelling rates Γ_1 and Γ_2, Equations 2.32 and 2.33, are very different and the current rises in a stepwise manner, referred to as the Coulomb staircase (Fig. 2.12). Each successive step in the Coulomb staircase corresponds to an electron added ($V > 0$) or removed ($V < 0$) from the island. For a double tunnel junction with equal tunnel junction capacitances $C_1 = C_2 = C$ and arbitrary values of R_1 and R_2, the current is given by (Ingold and Nazarov, 1991):

$$I(V_l) = \frac{e}{C(R_1 + R_2)} l \qquad (2.44a)$$

at specific values of voltages V_l given by:

$$V_l = \frac{e}{C}(l + \frac{1}{2}), \qquad l = 0, 1, 2... \qquad (2.44b)$$

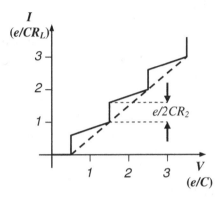

Fig. 2.12 The Coulomb staircase.

Thus, as l increases, I increases linearly. At $l = 0$, $V_l = e/2C$ is at the edge of the Coulomb gap (Fig. 2.11) and $I = 0$. Considering only the specific points $V_l = e/2C$, $3e/2C$, $5e/2C$, etc. separated by $V = e/C$, the current increases linearly. However, in between the points given by Eq. 2.44b, the characteristics depend on R_1 and R_2. Assuming $R_1 \ll R_2$, then at a positive voltage V greater than the Coulomb gap, the island charges up through the first junction to the maximum value possible, n_{max}, until the tunnel rate $\Gamma_{1,l\text{-}r}$ drops to zero. Using Equations 2.30a and 2.33a, this is given by:

$$n_{max} < -\frac{1}{2} + \frac{CV}{e} \qquad (2.45)$$

where n_{max} is an integer. As the tunnel rate through the second junction $\Gamma_{2,l\text{-}r}$ is smaller, whenever an electron tunnels off the island across the second junction, the first junction rapidly recharges the island. The current is then determined by the tunnelling rate of the second junction, which acts as the 'bottleneck'. Using Equations 2.30d and 2.33b, this is given by:

$$\Gamma_{2,l-r} = \frac{1}{2R_2C}\left(n_{max} - \frac{1}{2} + \frac{CV}{e}\right) \qquad (2.46)$$

The current through the double junction is then given by:

$$I(V) = e\Gamma_{2,l-r} = \frac{1}{2R_2C}\left(n_{max}e - \frac{e}{2} + CV\right) \qquad (2.47)$$

At voltages given by the points of Eq. 2.44b, and observing that $n_{max} = l$, Eq. 2.47 reduces to Eq. 2.44a. However, if we increase V just above one of the points of Eq. 2.44b, then n_{max} increases by 1, increasing I by $e/2R_2C$. This step in the current is followed by a linear increase in the current until the next point of Eq. 2.44b is reached. The net result is a stepwise increase in the current, creating the Coulomb staircase I-V characteristic shown by the solid line in Fig. 2.12. The step width is given by $\Delta V = e/C$. In contrast, if $R_1 \sim R_2$, then electrons tunnel off the island across the second junction at similar rates to electrons tunnelling onto the island across Junction 1, and the current increases linearly (dashed line, Fig. 2.12) between the points of Eq. 2.44b. Here, the Coulomb staircase is not observed, or is, at best, very faint. In the preceding analysis we assume equal capacitances – with unequal capacitances, the staircase may be more complicated.

Figure 2.13 shows a simulation of the I-V (Fig. 2.13[a]) and I-V_g (Fig. 2.13[b]) characteristics of an SET at $T = 4.2$ K. The simulation is carried out using a programme called 'CAMSET' (Amakawa et al., 1998), which utilizes a Monte Carlo simulation method. Here, we assume $C_1 = C_2 = C = 1$ aF, $C_g = 0.1$ aF, $R_1 = 500$ kΩ, $R_2 = 5$ MΩ and zero offset charge. In the I-V characteristics, as V_g is varied from 0 V to a value $\sim e/2C_g = 0.8$ V, the width of the Coulomb gap reduces from a maximum of $2 \times e/2C_1 = 0.16$ V to zero. Increasing V_g from -0.8 V to 0 V modulates the Coulomb gap from zero to the maximum value of 0.16 V. The resistance mismatch of $R_2/R_1 = 10$ leads to a strong Coulomb staircase, with a step width given by $\Delta V = 0.16$ V. In the I-V_g characteristics (Fig. 2.13[b]), Coulomb oscillations are seen in the current, increasing in magnitude as V increases. The period of the oscillations is $\Delta V_g = e/C_g = 1.6$ V. The oscillations are shown for V

varying from 0.2 V to 0.1 V, mainly within the Coulomb gap of 0.08 V. It is seen that the oscillations are asymmetric in shape, a consequence of the resistance mismatch (Ingold and Nazarov, 1992). If we choose equal resistances, e.g. $R_1 = R_2 = 2.5$ MΩ, (Fig. 2.13[c]), the oscillations become symmetric in shape.

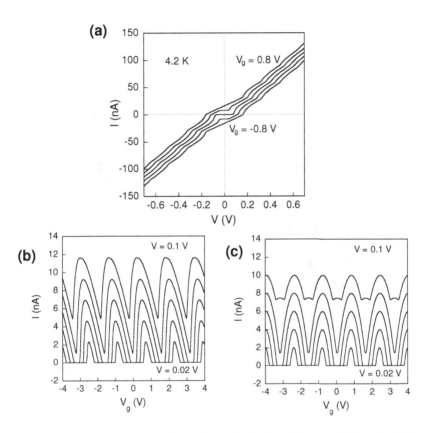

Fig. 2.13 (a) Monte Carlo simulation of the I-V characteristics, at 4.2 K, of a SET where $C_1 = C_2 = 1$ aF, $C_g = 0.1$ aF, $R_1 = 500$ kΩ and $R_2 = 5$ MΩ. (a) Current I vs. drain voltage V as the gate voltage V_g varies from −0.8 V to 0.8 V in 0.4 V steps. The curves are offset 10 nA from each other for clarity. (b) Coulomb oscillations in I vs. V_g as V varies from 0.02 V to 0.1 V in 0.05 V steps. (c) Coulomb oscillations for $R_1 = R_2 = 2.5$ MΩ, and C_1, C_2, C_g and V as in (b).

Figure 2.14(a) shows a 3-D plot of the simulated I-V characteristics of an SET at $T = 4.2$ K. We assume $C_1 = 1$ aF, $C_2 = 2$ aF, $C_g = 0.1$ aF, $R_1 =$

2 MΩ, R_2 = 1 MΩ and zero offset charge. Here, V_g is varied from −2.6 V to +2.8 V, a range which covers three complete 'Coulomb diamonds'. Figure 2.14(b) shows a grey-scale image of the SET conductance, $g = \delta I/\delta V$, as a function of V and V_g. The conductance g forms a sharp peak at each current step, corresponding to the values of V and V_g where a transition in the electron number n occurs. The plot traces out the stability diagram of the SET, in a manner similar to Fig. 2.11. Here, $C_1 \neq C_2$ and the Coulomb diamonds are 'tilted' with respect to the V axis, e.g. for the central Coulomb diamond where $n = 0$, the corners a and b do not lie on the line where $V_g = 0$ V.

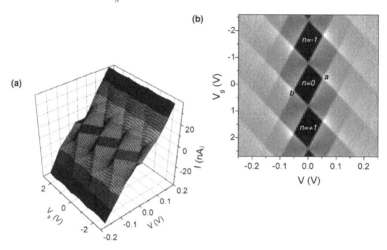

Fig. 2.14 Monte Carlo simulation of the I-V characteristics, at 4.2 K, of a SET where C_1 = 1 aF, C_2 = 2 aF, C_g = 0.1 aF, R_1 = 2 MΩ and R_2 = 1 MΩ. (a) 3-D plot of the current I vs. drain voltage V and gate voltage V_g. (b) Grey-scale plot of the conductance g vs. V and V_g. The conductance varies from 0 to 544 nS.

2.4.6 Energy band diagrams

In a manner similar to our analysis of the single-electron box, an energy band diagram can be drawn for the SET. However, in contrast to the single-electron box, in the SET the electron number n can be changed by varying two different voltages, the drain-source voltage V or the gate voltage V_g. Again, if the single-electron energy levels E_n lie below the Fermi energy, then n electrons exist on the island.

Initially assuming an SET at $V_g = 0$ V, where $C_g \ll C_1$ or C_2 for simplicity, the energies E_n may be calculated using Eq. 2.30a for a double tunnel junction. This gives the energy needed for an electron to be added to the island across Junction 1, with the initial island state n. For $V = 0$, $n = 0$ and assuming equal tunnel junction capacitances $C_1 = C_2 = C$, a total energy $\Delta E_{1,add} = e^2/2C_\Sigma = e^2/4C$ must be provided for an electron to be added to the island. This gives the position in energy where the first electron is added, $E_1 = e^2/4C$. Now, adding a second electron to the island is equivalent to adding one electron to the island in a state $n = 1$. For $n = 1$, Eq. 2.30a then gives $E_2 = 3e^2/2C_\Sigma = 3e^2/4C$. Similarly, $n = 2, 3...$ give the positions in energy E_3, E_4, etc., where the 3rd, 4th etc. electrons are added to the island. Equation 2.30b may be used to give the positions E_{-1}, E_{-2}, E_{-3}, etc. This allows us to draw the energy band diagram for the SET shown in Fig. 2.15(a). At $V = 0$, the states for $n > 0$ are empty, and the states for $n < 0$ are filled.

The energy band diagram provides a simple picture to understand the conduction mechanism across the SET. If a voltage $+V$ is applied to the drain, then a bias drops across the two tunnel junctions, lowering the positions in energy relative to the Fermi energy in the source E_{FS}. For tunnel junctions with equal capacitance C, half of V drops across each tunnel junction, and at a voltage $V_1 = e/2C$, E_1 is level with E_{FS} (Fig. 2.15[b]). At this point, an electron tunnels onto the island from the source, sequentially followed by tunnelling off the island into the drain. The process then repeats itself and a current begins to flow across the SET, with the average electron number given by $n = 1$. As discussed in the previous section, the current increases linearly if the tunnel resistances are similar (dashed line, Fig. 2.12), and in a stepwise manner if they are very different (solid line, Fig. 2.12[b]). As V is increased to $V_2 = 3e/2C$, E_2 becomes level with E_{FS}, causing n to increase to 2 (Fig. 2.15[c]). Further increases in V lead to increasing n, consistent with the stability diagram of Fig. 2.11. Note that the tunnel rates are such that all levels E_n lying below E_{FS} are filled. In this picture, E_n mark only the *energy* where the electron number changes and do not represent resonant tunnelling levels. A negative value of V reduces n.

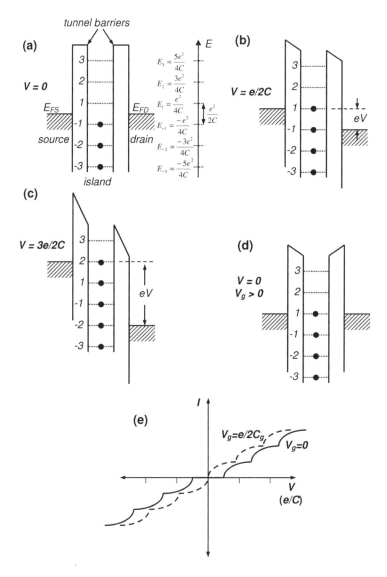

Fig. 2.15 Energy diagrams in a SET. (a) Energy diagram at drain-source voltage $V = 0$ and gate voltage $V_g = 0$. (b) At $V = e/2C$. (c) At $V = 3e/2C$. (d) At $V = 0$ and gate voltage $V_g > 0$. (e) Coulomb staircase for $V_g = 0$ and for $V_g = e/2C_g$.

Finally, we consider the effect of the gate bias V_g. If V_g is increased to positive values at $V = 0$ V, the energy of the levels E_n is changed by an amount depending on the capacitive divider formed by C_g, in series with the parallel combination of C_1 and C_2. The energy shift is given by $\Delta E = -eC_gV_g/(C_1 + C_2) = -eC_gV_g/2C$, as $C_1 = C_2 = C$. If $\Delta E = E_1$, corresponding to $C_gV_g = e/2$, the E_1 is aligned with E_{FS} and $n = 1$ (Fig. 2.15[d]). The Coulomb gap is then overcome even at $V = 0$ V and a current flows even for small V, consistent with the stability diagram of Fig. 2.11. Figure 2.15(e) shows schematically the position of the Coulomb staircase for $V_g = 0$ (solid line), and for $V_g = e/2C_g$ (dashed line). Electrons can be added or removed from the island by applying positive or negative values of V_g respectively.

2.5 Quantum Dots

In previous sections, we considered single-electron effects in metal systems, where a number of free electrons existed on the island. This discussion ignored the possibility that the island could be small enough such that the de Broglie wavelength of the electrons was comparable to the island dimensions. For small island sizes, free electrons within the island are confined by the surface potential, forming a 'particle-in-a-box'. The electrons then occupy discrete energy levels on the island, and if the energy level spacing $\Delta E \gg k_BT$, these levels effect the conduction mechanism across the island. The island is then referred to as a 'quantum dot' (QD) (Reed *et al.*, 1988; Reed, 1993; Kastner, 1993), and the conduction mechanism depends on a combination of quantum confinement and single-electron effects. The energy level spectrum on the quantum dot is discrete, and the quantum dot is analogous to an 'artificial atom' (Kastner, 1993), where the energy spectrum can be measured electrically.

A large body of work exists on quantum dots in many systems, perhaps the most popular of which are two-dimensional electron gas (2-DEG) systems in III-V heterostructure materials such as GaAs/AlGaAs. In this section, we will only provide a basic introduction to quantum dots. For more details, the reader is referred to various theoretical analyses of these systems (Averin *et al.*, 1991; Beenakker, 1991; Meir *et al.*, 1991), and review articles (van Houten *et al.*, 1991; Kouwenhoven *et* al., 1997; Ashori, 1996; Meirav and Foxman, 1996). A silicon QD SET is analysed by Natori et al. (Natori et al., 2000).

Consider electrons confined in a 3-D quantum dot, with sides of length L_x, L_y and L_z, along the x-, y- and z-axis. Assuming a hard-walled surface potential, the dot behaves as a 3-D potential well and simple quantum mechanics gives the energy levels in the box:

$$E_{n_x,n_y,n_z} = \frac{\hbar^2 \pi^2}{2m^*}\left(\frac{n_x^2}{L_x^2} + \frac{n_y^2}{L_y^2} + \frac{n_z^2}{L_z^2}\right) \quad (2.48)$$

where n_x, n_y and n_z are the quantum numbers along the x-, y- and z-axis and m^* is the effective mass. For sides of different length, the energy levels are non-degenerate (ignoring spin). For a cubic quantum dot, with equal sides of length L, one may rewrite this equation in terms of a single quantum number n:

$$E_n = \frac{\hbar^2 \pi^2 n^2}{2m^* L^2} \quad (2.49)$$

where $n^2 = n_x^2 + n_y^2 + n_z^2$. Here, the levels may be degenerate, e.g. for the level E_1, $n = 1$ corresponds to values of (n_x, n_y, n_z) of (1, 0, 0), (0, 1, 0) or (0, 0, 1), i.e. three-fold degenerate ignoring spin, or six-fold degenerate including spin.

In many cases, the quantum dot may have a parabolic confinement potential profile $V(x) = \frac{1}{2} kx^2$, where k is the equivalent of the 'spring constant' and determines the shape of the potential. In this case, the quantum dot is equivalent to a harmonic oscillator and the energy level spacing ΔE_s is constant. In a 1-D quantum dot of total width $2r$, where the barrier height is V_0 at r, using $\omega = \sqrt{(k/m^*)}$, the energy level spacing is:

$$\Delta E_s = \hbar \omega = \Delta E_s = \hbar \omega = \hbar\sqrt{\frac{2V_0}{m^* r^2}} \quad (2.50)$$

We may also obtain more generalized expressions for the energy level spacing, ΔE, at the Fermi energy for a 'box' of size L in 1-D, 2-D or 3-D. Including spin degeneracy, ΔE is given by (Kouwenhoven et al., 1997):

1-D $$\Delta E = \left(\frac{n}{4}\right)\frac{\hbar^2\pi^2}{2m^*L^2} \qquad (2.51a)$$

2-D $$\Delta E = \left(\frac{1}{\pi}\right)\frac{\hbar^2\pi^2}{2m^*L^2} \qquad (2.51b)$$

3-D $$\Delta E = \left(\frac{1}{3\pi^2 n}\right)\frac{\hbar^2\pi^2}{2m^*L^2} \qquad (2.51c)$$

We note that ΔE increases with n in 1-D, remains constant with n in 2-D, and decreases with n in 3-D. The energy scale of the level spacing is given by $\Delta E_s \sim \hbar^2\pi^2/m^*L^2$.

The dot size L in silicon-based single-electron devices can easily be ~10 nm, far smaller than the dot size in GaAs/AlGaAs 2-DEGs, where L ~100 nm. For a 3-D silicon quantum dot with $L = 10$ nm, using an average effective mass $m^* = 0.3m_0$ (Sze, 1981), Eq. 2.49 predicts that the first energy level lies at $E_1 = 12.5$ meV. This corresponds to the thermal energy k_BT at $T = 140$ K. This value of E_1 is far higher than the corresponding value for a 2-D dot in a GaAs/AlGaAs 2-DEGs, where E_1 ~0.6 meV, corresponding to the thermal energy at 7 K. Measurements on quantum dots in GaAs/AlGaAs 2-DEGs are often limited to milli-Kelvin temperatures. In contrast, the very small size scales possible in Si quantum dots raise the measurement temperature and for sub-5 nm dots, even room temperature measurement may be possible. Furthermore, the elastic mean free path of the electrons in the dot may need to be larger than the dot dimensions, in order for the size of the dot to determine the energy levels. In GaAs/AlGaAs 2-DEGs at low temperatures, the elastic mean free path may be very long, ~10 μm or even greater, and the typical experimental dot sizes of ~100 nm are far smaller than this. However, in a Si crystal, the mean free path may be very small, ~10 nm (Fischetti and Laux, 1988). In order for the energy levels to be determined by the quantum dot dimensions, the quantum dot should be ~10 nm or less in size.

2.5.1 Coulomb oscillations in quantum dots

We now discuss Coulomb oscillations in a quantum dot (Averin et al., 1991; Beenakker, 1991; Meir et al., 1991; Kouwenhoven et al., 1997). For simplicity, we assume a constant energy level spacing ΔE_s (Fig. 2.16), which may be calculated from Eq. 2.50. In quantum dots where a vertical, hard-walled potential is more realistic, ΔE_s varies with the quantum number and may be calculated from Eq. 2.48. We assume that the dot size is small enough such that the energy level spacing ΔE_s and the Coulomb charging energy E_c are both greater than the thermal energy $k_B T$, and that $\Delta E_s < E_c$. We also assume that the energy levels may be determined independently of the number of electrons on the dot. A more accurate analysis requires a self-consistent calculation of the energy levels, where the number of electrons on the dot and the dot potential are interrelated (McEuan et al., 1993). Finally, we assume that a constant capacitance $C_\Sigma = C_1 + C_2 + C_g$ is associated with the dot.

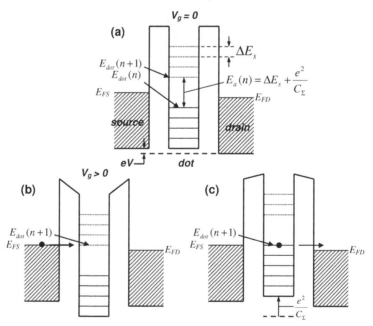

Fig. 2.16 Energy band diagrams for a quantum dot. (a) At $V_g = 0$ and a small drain-source voltage V. (b) At $V_g > 0$, before an electron tunnels onto the dot. (c) At $V_g > 0$, after an electron tunnels onto the dot.

A schematic diagram of the quantum dot is shown in Fig. 2.16, following the picture given by Kouwenhoven *et al.* (Kouwenhoven *et al.*, 1991a). We assume metallic contacts, where the Fermi energies of the source and drain are E_{FS} and E_{FD}, measured relative to the bottom of the conduction band. We also assume that the applied bias V is low, i.e. smaller than $\Delta E_s/e$ or e/C, the linear-response regime. We then have $E_{FS} \approx E_{FD}$, and electrons fill the energy levels in the dot below E_{FS}. The electrochemical potential of the dot, E_{dot}, for n extra electrons on the dot at a gate voltage V_g, is given by (Kouwenhoven *et al.*, 1991a):

$$E_{dot}(n) = E_n + \frac{\left(ne - \frac{1}{2}\right)e}{C_\Sigma} - \frac{C_g V_g}{C_\Sigma}e \qquad (2.52)$$

Here, E_n is the energy of the nth quantum confinement level, and the second two terms correspond to the electrostatic charging energy necessary to add n electrons to the dot by means of the gate voltage. E_n corresponds to the highest value of n for which $E_n < E_{FS} \approx E_{FD}$. We note that the charging energy necessary to add an electron to the dot, given n electrons on the dot, obtained from Eq. 2.52 by substituting $n + 1$ for n and ignoring E_n, is consistent with Eq. 2.28a with $V = 0$.

The 'addition energy' E_a is the change in the electrochemical energy of the dot for one electron added to the dot at a fixed gate voltage, $E_{dot}(n+1) - E_{dot}(n)$. Using Eq. 2.52, this is given by:

$$E_a(n) = E_{dot}(n+1) - E_{dot}(n) = \Delta E_s + \frac{e^2}{C_\Sigma} \qquad (2.53)$$

In this model, the charging energy term e^2/C_Σ manifests only at the Fermi energy, as shown in Fig. 2.16(a). Below the Fermi energy, the electron states are separated only by the quantum level spacing ΔE_s, also referred to as the 'excitation energy' of the quantum dot. The addition energy leads to an energy gap at the Fermi energy, leading to Coulomb blockade of the dot. The Coulomb blockade may be overcome by using the gate voltage, as discussed earlier in the discussion of the SET. If V_g is increased, $E_{dot}(n+1)$ is reduced until it aligns with E_{FS} (Fig. 2.16[b]). An electron then tunnels onto the dot, increasing the electrostatic energy of the dot by e^2/C_Σ (Fig. 2.16[c]). In this picture, there is an increase in the bottom of the conduction band in the dot corresponding to e^2/C_Σ. However, as $E_{dot}(n+1) > E_{FD}$, the electron tunnels off the dot. This process then repeats and a current flows across the quantum dot. As V_g is swept further, Coulomb oscillations are observed as the Coulomb blockade is overcome successively and the dot charges with extra electrons. Using Eq. 2.52 and the condition $E_{dot}(n+1, V_g + \Delta V_g) = E_{dot}(n, V_g)$, we can obtain the period of the Coulomb oscillations:

$$\Delta V_g = \frac{C_\Sigma}{eC_g}\left(\Delta E_s + \frac{e^2}{C_\Sigma}\right) \qquad (2.54)$$

Equation 2.54 may be compared to the period of the Coulomb oscillations in a metallic SET, $\Delta V_{g,SET} = e/C_g$. In contrast to an SET, in a quantum dot, *aperiodic* Coulomb oscillations are observed, though the level of aperiodicity may be small if $\Delta E_s \ll e^2/C_\Sigma$. Considering spin, a second electron of opposite spin may be added to an energy level. Furthermore, often ΔE_s is not constant and varies from level to level, increasing the aperiodicity of the Coulomb oscillations.

Finally, we briefly mention changes in the height of the Coulomb oscillation peaks. As each peak corresponds to a specific energy level, and the tunnelling probability depends on the shape of the electron wavefunction of the level, in a quantum dot the Coulomb oscillation peaks can vary in height in a complex manner. This is unlike a metallic SET, where peaks of similar height are observed. However, irregular peak heights do not, by themselves, confirm the presence of quantum confinement. Any variation in the tunnel barrier with gate bias would also create such an effect. This possibility cannot be ignored in many Si SET devices, where the tunnel barrier is formed by depleted regions of Si which are created by the gate potential.

2.6 The Multiple-Tunnel Junction

The multiple-tunnel junction (MTJ) consists of a 1-D chain of nanoscale islands and tunnel junctions (Fig. 2.17). In such a system, the single-electron charging of each island, and the effect of excess electrons on the polarization of neighbouring tunnel junctions, modifies the Coulomb blockade region, and the *I-V* characteristics of the system (Amman *et al.*, 1989; Bakhvalov *et al.*, 1989; Ben-Jacob *et al.*, 1989; Likharev *et al.*, 1989). This can lead to the time- and space-correlation of tunnel events in the MTJ, and the formation of single-electron charge 'solitons' along the MTJ. Furthermore, the single-electron charging energy of an island embedded within a chain of islands and tunnel barriers is also increased somewhat. This raises the maximum temperature where single-electron effects are observed, in comparison with a simple double tunnel junction of similar island capacitance. The MTJ is of particular interest for single-electron devices in silicon, as these devices are either designed with multiple islands, or the fabrication process and material morphology naturally forms MTJs (Chapter 3). In this section, we will concentrate on the DC *I-V* characteristics of MTJs. The fabrication and experimental characteristics of Si MTJs will be discussed in the following chapters. For a more detailed review of charge soliton transport, and the effect of the time-correlation of tunnel events on the *I-V* characteristics, the reader is referred to the article by Delsing (Delsing, 1992).

Figure 2.17(a) shows the circuit diagram for an MTJ with N islands, and $N+1$ tunnel junctions. The islands are separated from each other, and from the source and drain regions at the ends, by tunnel junctions with capacitance C. The remaining capacitance of each island is represented by C_0, and a bias V is used to inject electrons into the MTJ. Here, we will consider the simple approximation of a long, homogenous MTJ (Likharev *et al.*, 1989; Delsing, 1992), which may be regarded as a half-infinite array of capacitors on either side of an island embedded within the MTJ. Alternatively, the MTJ may be investigated by numerically solving equations for the junction and stray capacitor voltages, and hence the tunnelling rates (Amman *et al.*, 1989; Ben-Jacob *et al.*, 1989), or by obtaining an exact analytic solution for these equations, for finite MTJs (Hu *et al.*, 1993).

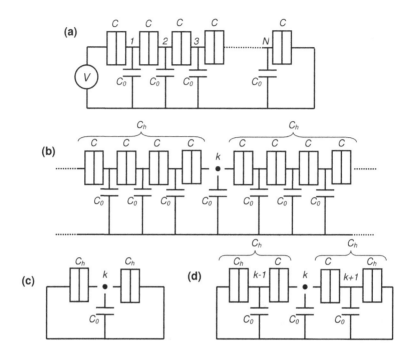

Fig. 2.17 (a) MTJ with N islands, $N+1$ tunnel capacitances C, and N stray capacitances C_0. (b) Looking from an island k, a long, homogenous MTJ can be approximated by two half-infinite capacitor arrays C_h. (c) Effective capacitance at island k is given by $C_0 + 2C_h$. (d) Looking from k, C_h is also given by a combination of capacitors, where C_h and C_0 are in parallel (at islands $k-1$ and $k+1$), and then in series with C.

We will now consider the capacitance of an island k within the MTJ (Fig. 2.17[b]). Looking from the island, we approximate each half of the MTJ by a half-infinite capacitor array, with capacitance C_h. The island capacitance may then be represented by an 'effective capacitance' $C_{eff} = C_0 + 2C_h$ (Fig. 2.17[c]). The potential v_k generated at the island by a single electron on the island is then given by $v_k = -e/C_{eff}$. It is straightforward to obtain an expression for C_h. Looking from the island, C_h is also given by a combination of capacitors, where C_h and C_0 are in parallel, and then in series with C (Fig. 2.17[d]). This gives:

$$\frac{1}{C_h} = \frac{1}{C_h + C_0} + \frac{1}{C}$$

$$\Rightarrow C_h = \frac{1}{2}\left(\sqrt{C_0^2 + 4CC_0} - C_0\right) \qquad (2.55)$$

and $C_{eff} = C_0 + 2C_h = \sqrt{C_0^2 + 4CC_0}$ \qquad (2.56)

The effective capacitance may be used to obtain the charging energy for a single electron on an island within the MTJ:

$$E_{cMTJ} = \frac{e^2}{2C_{eff}} \qquad (2.57)$$

Figure 2.17(d) may be used to calculate the potential of the $k-1^{th}$ island, when an electron is added to the kth island. For the kth island, the potential is given by $v_k = -e/C_{eff}$. This may be used to write the potential at the $k-1^{th}$ island:

$$v_{k-1} = -\frac{e}{C_{eff}}\left(\frac{C}{C + C_0 + C_h}\right) \qquad (2.58)$$

Repeating this process for successive islands, and for islands lying after the kth island, gives us the potential of the lth island, lying l number of islands away from the kth island:

$$v_l = -\frac{e}{C_{eff}}\left(\frac{C}{C + C_0 + C_h}\right)^{|k-l|} \qquad (2.59)$$

Figure 2.18 shows a plot of Eq. 2.59, for $C_0 = C/2$. It is seen that v_l decays as we move away from the kth island. This is associated with the reduction in the polarization of the capacitors on either side of the kth island as we move away from it. However, if $C_0 \approx 0$, i.e. the stray capacitance is negligible, we have all junctions charged the same magnitude.

Eq. 2.58 has been expressed in an exponential form (Likharev et al., 1989; Delsing, 1992):

$$v_l = -\frac{e}{C_{eff}} e^{-|k-l|/M} \quad (2.60)$$

$$\text{where } M = \left(\ln\left(\frac{C_{eff} + C_0}{C_{eff} - C_0} \right) \right)^{-1} \quad (2.61)$$

Here, M is the characteristic length of the potential distribution given by Eq. 2.60. For the plot of Fig. 2.18, the soliton is rather short, with $M = 1.44$.

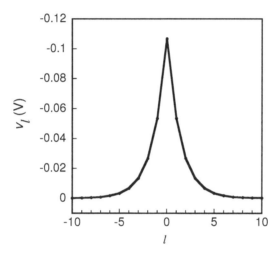

Fig. 2.18 Potential v_l of the lth island in a long, homogenous MTJ, lying l number of islands away from island k, which lies far from the edges. One electron is added to the island k. The tunnel capacitances $C = 1$ aF, and the stray capacitances $C_0 = 0.5$ aF. The potential distribution forms a charge soliton.

For a long MTJ with constant values of C and C_0, away from the edges of the MTJ, the potential distribution given by Equations 2.59 or 2.60 does not change in form, no matter which island the extra electron is placed on. Furthermore, as an electron tunnels from one junction to the next, it carries the potential distribution with it, i.e. a potential similar to Fig. 2.18 moves through the MTJ. The potential distribution is then referred to as a 'charge soliton'. It is possible to consider charge transport through the MTJ using a charge soliton picture (Amman *et al.*, 1989; Bakhvalov *et al.*, 1989; Ben-Jacob *et al.*, 1989; Likharev *et al.*, 1989). In this picture, the removal of an electron from an island forms a charge anti-soliton. Two charge solitons repel each other, and a soliton and anti-soliton attract each other. An anti-soliton and soliton will annihilate each other when they meet.

Furthermore, unbiased edges may be regarded as 'mirrors', such that a soliton at an electrode l near the edge acts as if the array was infinite and an image anti-soliton existed at an electrode $-l$. This implies that a soliton (or an anti-soliton), are attracted to the unbiased edge. Applying a large enough voltage V to an edge will inject a soliton into the MTJ, and a current begins to flow. The threshold voltage when this occurs is given by (Likharev *et al.*, 1989, Delsing, 1992):

$$V_t = \frac{e}{2C_{eff}}\left(1 + e^{-1/M}\right) = \frac{e}{C_{eff} + C_0} \quad (2.62)$$

The value of the threshold voltage is low compared to the voltage corresponding to the sum of the individual charging energies of the islands. This is because the applied voltage drops more across the first junction, closest to the voltage source, due to the presence of the stray capacitance C_0 at the first island (Amman *et al.*, 1989). As the voltage increases, the current increases such that for its asymptotic value, an offset in voltage is observed. This is given by:

$$V_{offset} \approx \frac{Ne}{2C} \quad (2.63)$$

For an MTJ where C_0 is negligible, V_t increases to a value close to V_{offset}.

In the soliton picture, the repulsion between a biased edge and the soliton pushes the soliton further into the MTJ, followed by a second soliton entering the MTJ. Ultimately, a train of solitons moves through the MTJ. This picture leads to the time-correlation of tunnel events within the MTJ. If the MTJ is irradiated with microwaves of frequency f, then phase-locking of the microwaves with the single-electron tunnelling oscillations leads to steps in the *I-V* characteristics at current $I = nef$. For an ideal MTJ system biased by equal voltages applied to either end, the solitons would meet in the middle, forming a stationary quasi-1D Wigner lattice (Likharev *et al.*, 1989). For unequal voltages, the Wigner lattice moves along the MTJ. A detailed discussion of these effects may be found in the articles by Bakhvalov *et al.* and Likharev *et al.* (Bakhvalov *et al.*, 1989; Likharev *et al.*, 1989).

Experimentally, MTJs with constant values of C and C_0 have been fabricated mainly using metal islands (Delsing *et al.*, 1989a; Kuzmin *et al.*, 1989). It is possible, at low temperatures, to observe the time-correlation of tunnelling events, and the phase-locking of the single-electron tunnelling oscillations to external microwave irradiation (Delsing *et al.*, 1989b, Delsing *et al.*, 1990). In contrast, in many silicon-based MTJs, the MTJ is random, with variation in the values of C and C_0 (Chapter 3). While, even in this case, a decaying potential distribution of the form of Fig. 2.18 may exist, this will change in form as it moves through the MTJ and cannot strictly be regarded as a charge soliton. However, it is possible to extend the method of image solitons, developed to describe the behaviour of solitons at the ends of homogenous MTJs, to obtain analytic solutions for homogenous and inhomogeneous finite MTJs (Jalil and Wagner, 1999).

We will now consider the *I-V* characteristics of the MTJ. As the bias V increases, the number of transitions in the MTJ increases. To understand this, following the argument of Amman *et al.* (Amman *et al.*, 1989), we may consider a simple two island MTJ, where the electron numbers on the islands are n_1 and n_2 respectively. Alternatively, we may consider a longer MTJ which holds two solitons at a time along its length, defined by n_1 and n_2 respectively. The charge state of the MTJ without any extra electrons (or solitons) may be expressed as $(n_1, n_2) = (0, 0)$. When a bias $V > V_t$ is applied, an electron tunnels onto the first island, creating the state (1, 0). This electron can then move through the MTJ, through the states (0, 1) to (0, 0) as it leaves the MTJ. However, as the applied voltage is increased, other transitions become possible, e.g. one may transition from the state (0, 1) either to the state (0, 0) or to the state (1, 1). These additional transitions lead to an increase in the current. For a longer MTJ, larger numbers of transitions are possible. If an asymmetry or non-homogeneity exists in the MTJ, e.g. if the tunnel resistances are different or the MTJ is driven asymmetrically by a voltage at one end only, then with each additional transition, a step in the current is observed. This leads to a Coulomb staircase at low voltages above V_t (Amman et al., 1989; Kuzmin et al., 1989). The staircase is typically irregular, with variation in the step heights and widths. As discussed earlier, the staircase tends asymptotically to a line offset in voltage by a value V_{offset} (Eq. 2.63).

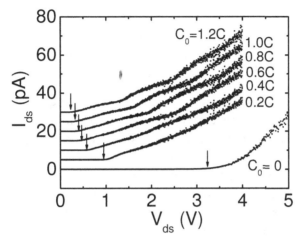

Fig. 2.19 Single-electron Monte Carlo simulation of the I-V characteristics of an MTJ at 300 K, with $N = 7$, $C = 0.12$ aF and $C_0 = 0.12$ aF. The tunnel resistance varies randomly by 60%. The average tunnel resistance of the MTJ, $R_{av} = 6$ GΩ.

Figure 2.19 shows a single-electron Monte Carlo simulation of the I-V characteristics of an MTJ at a temperature of 300 K, with $N = 7$, $C = 0.12$ aF and $C_0 = 0.12$ aF. These values are similar to those in an MTJ capable of room-temperature operation, fabricated using a 1-D chain of Si nanocrystals (Rafiq et al., 2008). A random variation in tunnel resistance of 60% was used to obtain a Coulomb staircase. The average tunnel resistance of the MTJ, $R_{av} = 6$ GΩ. The figure shows the effect of increasing C_0 from 0 to $1.2C$. As C_0 is increased, V_t reduces (arrowed), and the clarity of the staircase improves. For $C_0 = 0$, a high value of $V_t \sim$ 4 V is observed. This value tends towards the offset voltage of the MTJ, i.e. $V_t \sim V_{offset} = Ne/2C = 4.7$ V.

Chapter 3

Single-Electron Transistors in Silicon

3.1 Early Observations

The development of sub-micron lithographic techniques in the 1980s, and the reduction of the channel dimensions of MOSFETs to below 1 μm, raised the possibility of the observation of mesoscopic effects in semiconductor devices. In measurements of the low-temperature conductance of Si MOSFETs with 1-D channels, reproducible universal conductance fluctuations (UCF) (Alt'shuler, 1985; Lee and Stone, 1985) were observed as a function of the gate voltage, as the Fermi energy in the device was changed near the threshold voltage. These conductance oscillations were of order e^2/h, independent of the device size and geometry, and random in period. In 1989, Scott-Thomas *et al.* (Scott-Thomas *et al.*, 1988, 1989) observed *periodic* oscillation in the conductance of a 1-D channel in a Si MOSFET at low temperature. The channel in this device was formed using a stacked dual gate arrangement, where a split-gate with a 70 nm gap was defined between the silicon surface and a large upper gate. An electric field, created using the upper gate, was used to define a ~30 nm channel in the inversion layer (Fig. 3.1[a–b]). Scott-Thomas *et al.* suggested that these oscillations could be explained by the pinning of a charge density wave (CDW) or 'Wigner lattice' (see Grüner, 1988) by scattering centres associated with interface charges. In this model, the interaction between electrons in the 1-D channel and the underlying lattice potential caused a periodic variation in the charge of the 1-D channel, the CDW. The CDW could be pinned by the potential of two or more interface charges. For charges separated by a distance L_0, the pinning energy was minimized when the mean density corresponded to an integer number of electrons n (per unit length of channel) between the two pinning centres. Varying the gate potential changed n and led to periodic oscillations in a plot of conductance vs. n, with the period proportional to $1/L_0$.

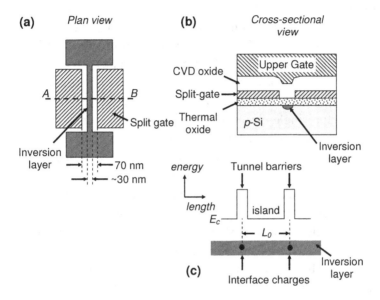

Fig. 3.1 Split-gate narrow channel MOSFET (Scott-Thomas *et al.*, 1989). (a) Plan view. (b) Cross-sectional view, through dashed line A–B in the plan view. (c) Edge of conduction band E_c, shown along the inversion layer. Two interface charges lie in the inversion layer.

The CDW picture is not the only means of understanding periodic conductance oscillations in a 1-D channel. Van Houten and Beenaker (van Houten and Beenaker, 1989) suggested that single-electron charging effects in the 1-D channel could explain the observations of Scott-Thomas *et al.*. Interface charges, or an impurity potential along the channel, could form two or more tunnel barriers along the channel with a charging island confined between the barriers (Fig. 3.1[c]). In such a system, for a total island capacitance C and island-to-gate capacitance C_g, the single-electron charging energy is $e^2/2C$ and Coulomb oscillations of period e/C_g occur as a function of gate bias. We note that, as C and L_0 vary in a similar manner in this geometry, it can be difficult to distinguish between the two mechanisms (Kastner *et al.*, 1989, Field *et al.*, 1990).

Single-electron charging effects in 1-D channel MOSFETs at low temperature are a direct consequence of the reduction in the MOSFET dimensions. As device size is reduced into the nanometre scale, a combination of the reduction in the total channel capacitance and the presence of disorder (in the above case, the disordered distribution of interface charges) can easily lead to the channel breaking up into nanoscale islands isolated by tunnel junctions, with capacitances small enough for single-electron charging. However, in order to design devices specifically for single-electron operation, configurations other than a MOSFET configuration can be more appropriate. For example, high-resolution lithography can be used to directly define the charging island between source and drain regions. A full SET can be defined by the addition of a gate electrode near the island, allowing the use of a gate voltage to electrostatically control the island charge and device current.

Silicon single-electron devices with lithographically defined islands were first fabricated by Paul *et al.* (Paul *et al.*, 1993) in a SiGe δ-doped layer, and by Ali and Ahmed (Ali and Ahmed, 1994) in SOI material. In the later device, full SET operation was demonstrated. Nakajima *et al.* (Nakajima *et al.*, 1994) fabricated doped nanowires in SOI material and observed periodic oscillations in the device current with gate voltage in a manner analogous to the work of Scott-Thomas *et al.*, although this was a doped system rather than an inversion layer. Here, it was also possible to attribute these oscillations to single-electron effects.

These initial devices were followed soon after by a dramatic improvement in the state of the art, with the fabrication of the first single-electron devices operating at room temperature (Takahashi *et al.*, 1995; Yano *et al.*, 1995). These devices used ultra-small islands ~10 nm or less in diameter, defined either by a combination of high-resolution electron-beam (e-beam) lithography and pattern-dependant oxidation (PADOX) (Takahashi *et al.*, 1995), or 'naturally' by silicon nanocrystals in an ultra-thin (<5 nm) nanocrystalline silicon film (Yano *et al.*, 1995).

This chapter describes the design, fabrication and characterization of SETs in silicon material. The first part of the chapter, Section 3.2, discusses SETs fabricated in crystalline silicon. This is then followed by a discussion of SETs fabricated in nanocrystalline silicon materials (Section 3.3). The final part of the chapter, Section 3.4, discusses single-electron effects in silicon nanowires and nanochains synthesized by material growth processes rather than by lithographic techniques.

3.2 SETs in Crystalline Silicon

A wide variety of SETs have been demonstrated in crystalline silicon, using many different techniques to define the nanoscale tunnel barriers and charging islands. Typically, the charging islands can either be defined using high-resolution lithographic methods, or by using the disorder inherent at the nanometre-scale in a doped silicon nanowire. Both single-electron and single-hole transistors have been demonstrated. In the following, we discuss these two different approaches to the design and fabrication of crystalline silicon SETs. Section 3.2.1 discusses SETs with islands defined using high-resolution lithography and Section 3.2.2 discusses SETs based on MOSFET structures. Section 3.2.3 discusses SETs using silicon nanowires. We then discuss SETs where the single-electron oscillations have large peak-to-valley ratios, even at room temperature (Section 3.2.4). The final part of this section discusses the fabrication and characterization of a nanowire SET in SOI material (Section 3.2.5).

3.2.1 SETs with lithographically defined islands

High-resolution lithographic techniques such as e-beam lithography and reactive-ion etching (RIE) can be used to define the charging island of an SET. Usually, oxidation of the silicon is also carried out, to further reduce the island dimensions and to passivate the surface and defect states created during the etching process. This approach provides a means to define an island of known dimension accurately. It is then theoretically possible to design the device to obtain a required Coulomb gap, and to obtain better reproducibility between the electrical characteristics of different devices – essential requirements for the development of large-scale single-electron circuits.

3.2.1.1 Etched islands

Most demonstrations of silicon SETs with lithographically-defined islands are in SOI material, either the separation by implantation of oxygen (SIMOX) type, the bonded oxide type (Colinge, 1997) or SiGe material (Cain *et al.*, 2001). In most devices in SOI material, a thin top silicon layer ~50 nm or less is used. The top silicon layer may be doped *n*-type (Ali and Ahmed, 1994) or *p*-type (Leobandung *et al.*, 1995a) to obtain either single-electron or single-hole operation. Trench isolation techniques can be used to define a planar device in the top silicon layer. A typical device design is shown schematically in Fig. 3.2(a). Here, an approximately cylindrical island is connected by narrower silicon constrictions or 'necks' to source and drain contact regions of larger area (Ali and Ahmed, 1994; Leobandung *et al.*, 1995a, 1995b, 1995c; Köster *et al.*, 1997; Koester, *et al.*, 1998; Sakamoto *et al.*, 1998; Zhuang *et al.*, 1998; Augke *et al.*, 2000; Rokhinson *et al.*, 2000).

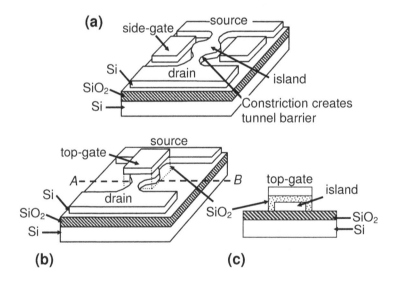

Fig. 3.2 (a) SET in SOI material using an etched island and side-gates. (b) SET in SOI material, with top-gate. (c) Cross-sectional view along dashed line *A–B*.

It is also possible to fabricate double islands (Cain *et al.*, 2001a, 2001b), or chains of islands separated by necks (Nakajima, 2002a). The entire pattern can be defined in the top silicon by RIE, where trenches isolate the various parts of the device from each other. The surface potential in the necks is greater than in the somewhat larger island, leading to potential barriers in the necks. The potential barriers then act as tunnel barriers, coupling the island to the source and drain. The gate electrode of the device may also be formed in a variety of ways. Perhaps the simplest technique is to use two in-plane gates, also in the top silicon of the SOI material, and on either side of the island (Fig. 3.2[a]) (Köster *et al.*, 1997; Koester *et al.*, 1998; Sakamoto *et al.*, 1998; Augke *et al.*, 2000; Cain *et al.*, 2000, 2001; Nakajima *et al.*, 2002a).

These 'side-gates' can be defined in the same lithographic pattern as the island and tunnel barriers, simplifying the fabrication process. With good etching of the trench between the gate and island down to the buried oxide (BOX) layer, this approach produces good gate isolation and low gate leakage currents. Alternatively, a top-gate may be defined above the island, formed by a gate stack similar to that in a MOSFET (Fig. 3.2[b–c]). This approach has the advantage of a thin gate oxide, better gate control of the island charge and a higher transconductance (Leobandung *et al.*, 1995a, 1995b, 1995c; Zhuang *et al.*, 1998; Rokhinson *et al.*, 2000). The gate may also be supported on an oxide or nitride layer formed by chemical vapour deposition (CVD).

Finally, the substrate of the SOI material may be used as a back-gate, where the gate electric field acts across the BOX layer (Fig. 3.2[c]). As the back-gate exists inherently in an SOI device, it may also be used in conjunction with other gates very conveniently, e.g. in combination with side-gates (Augke *et al.*, 2000). However, the thickness of the BOX (~40 nm to ~100 nm) is usually much greater than the top-gate oxide, implying that larger back-gate voltages are necessary to control the device. There are also likely to be a greater number of defects within the BOX layer compared to a good gate oxide. This implies that the back-gate is less effective than the top-gate, especially in SETs for circuit applications.

The definition of the device by RIE also leads to a high density of defect states on the etched surfaces. This usually implies that it is necessary to passivate the defect states by thermal oxidation. The process also reduces the defect density at the Si-SiO$_2$ interface to $\sim 10^{10}$–10^{11}/cm^2 (Sze, 2002). For these defect densities, in an SOI island with a surface area 20 nm × 20 nm and a top-gate or a back-gate, there is likely to be no defect, or only a single defect at the Si-SiO$_2$ interfaces at the gate oxide. This greatly reduces the likelihood of 'offset' charge switching caused by interface charges influencing the device characteristics.

We will now consider the single-island SET design of Ali and Ahmed in detail (Ali and Ahmed, 1994). Figure 3.3 shows a schematic diagram and scanning electron microscope (SEM) image of the SET. The device was fabricated in SIMOX SOI material with a top silicon layer 50 nm in thickness, heavily-doped n-type by phosphorous implantation to 10^{14}/cm^2. This layer was then thinned by oxidation to 40 nm. From Hall measurements, the carrier concentration and mobility at 4.2 K was determined to be 8.8×10^{18}/cm^3 and 180 cm^2/V.s, respectively. The device was fabricated using a combination of optical lithography for the contacts, and e-beam lithography for the island and the connecting leads, as follows.

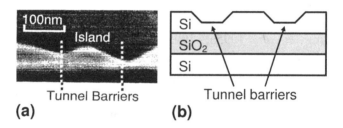

Fig. 3.3 (a) Scanning electron micrograph of a SET in SOI. (b) Schematic diagram. (Reprinted with permission from Ali and Ahmed, *Appl. Phys. Lett.* **64**, 2119 [1994]. Copyright 1994, American Institute of Physics).

An initial stage of e-beam lithography was used to define the patterns for a 1 μm wire with side-gates, in 400 nm-thick polymethylmethacrylate (PMMA) resist. The pattern was then transferred into the top silicon layer using RIE in a 1:1 ratio of SiCl$_4$ and CF$_4$, both at a flow rate of 20 sccm. The etching process used an r.f. power and frequency of 300 W and 13.56 MHz, respectively, and a chamber pressure of 20 mbar.

The wires were etched down to the BOX layer of the SOI material, to allow isolation from the side-gates. A second stage of e-beam lithography and RIE was then used to define two narrow regions along the wire, where the silicon was thinned to form tunnel barriers. The two tunnel barriers isolated a ~100 nm long island in between (Fig. 3.3[b]). The island was separated from the side-gates by ~500 nm.

The *I-V* characteristics of this device at 0.3 K are shown in Fig. 3.4. A Coulomb gap $V_c \approx 1.6$ mV was observed, symmetrical about zero bias. This corresponded to a total island capacitance $C_T = 50$ aF, estimated from $V_c = e/2C$. The characteristics were linear outside the Coulomb gap and a staircase was not observed, suggesting symmetrical tunnel barriers. The Coulomb gap was almost completely washed out thermally at only 3.8 K. The Coulomb oscillation period $\Delta V_G \approx 1$ V, corresponding to a gate capacitance $C_G = e/\Delta V_G = 0.16$ aF.

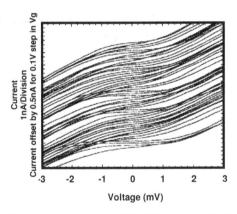

Fig. 3.4 *I-V* characteristics of the SET of Fig. 3.3, at 0.3 K. Each curve is offset by 0.5 nA/0.1 V step in gate voltage. (Reprinted with permission from Ali and Ahmed, *Appl. Phys. Lett.* **64**, 2119 [1994]. Copyright 1994, American Institute of Physics).

The very small Coulomb gap in this device, and the correspondingly low temperature for the observation of single-electron charging, is a consequence of the rather large island size of 100 nm and tunnel barrier width of ~30–50 nm. A reduction in the island size, and in the general device dimensions, is necessary to reduce the total island capacitance. This would increase the width of the Coulomb gap and lead to an increase in the operating temperature of the device. For ultra-small island sizes ~10 nm or less, the total island capacitance $C_T \sim 1$ aF and the charging energy $E_c = e^2/2C = 80$ meV, greater than $k_BT \approx 26$ meV at $T = 300$ K. SET operation then becomes possible, even at room temperature. For example, Zhuang *et al.* (Zhuang *et al.*, 1998) have fabricated an SET with multiple islands, where Coulomb oscillations are observed in the drain current vs. gate voltage (I_d-V_g) characteristics at room temperature. A schematic of this device is shown in Fig. 3.5. Here, the smallest island size was estimated to be only ~12 nm. The device used a 30 nm nanowire, fabricated by e-beam lithography and RIE in SOI material, with a ~30 nm-thick top silicon layer. Oxidation of the wire at 950°C reduced the wire dimensions to only 16 nm. Noise in the e-beam lithography led to fluctuations in the fabricated wire width, defining islands isolated by narrower tunnel barrier constrictions. At room temperature, only the smallest islands, ~12 nm in size, contributed to the characteristics. The device used a polysilicon top-gate, supported on a 25 nm-thick gate oxide layer deposited by plasma-enhanced chemical vapour deposition (PECVD) on the nanowire. From the charge stability 'Coulomb diamond' regions, seen as a function of the gate and drain voltages, the energy level separation was estimated to be ~130 meV. This is well above $k_BT \approx 26$ meV at room temperature, and leads to room temperature observation of the Coulomb oscillations.

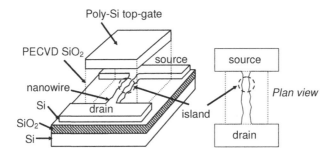

Fig. 3.5 Schematic diagram of the SET fabricated by Zhuang *et al.* (Zhuang *et al.*, 1998), capable of room temperature operation. An island only ~12 nm in size is defined by noise in the *e*-beam lithography.

Electron-beam lithography and RIE provide a very flexible means to fabricate SETs in various configurations, e.g. with one or more than one etched island. With good high-resolution e-beam lithography, it is possible to define both the islands and the tunnel barriers in the same stage of lithography. Cain *et al.* (Cain *et al.*, 2000, 2001) have fabricated a double island device in SiGe material by e-beam lithography and RIE, where the physical island size was ~50 nm and the tunnel barrier constrictions were only ~20 nm in size. Figure 3.6 shows an SEM image of the device. The device was defined in a 40 nm-thick layer of $Si_{0.9}Ge_{0.1}$ grown on an un-doped SOI wafer. The SiGe layer was heavily-doped *p*-type with boron, at a concentration of $10^{19}/cm^3$. Two side-gates were used, with a side-gate-to-island separation of 100 nm. The surface depletion in the material was ~10 nm, leading to an electrical island size of ~30 nm, isolated by fully depleted constrictions. The constrictions then formed very effective tunnel barriers. The dimensions of the islands and constrictions were carefully chosen and optimized to obtain un-depleted islands and fully depleted constrictions. Additional multiple-tunnel junctions (MTJs) were not formed in this device, leading to well-characterized electrical behaviour. This required good control over the surface depletion length, obtained by optimizing the doping density, device dimensions and the surface treatment.

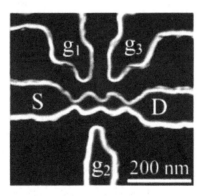

Fig. 3.6 SEM image of a double dot SiGe SET. (Reprinted with permission from Cain *et al.*, *Appl. Phys. Lett.* **78**, 3624 [2001]. Copyright 2001, American Institute of Physics).

In this device at 4.2 K, a Coulomb staircase with well-defined steps and a Coulomb gap of 60 meV was observed. The gate-island capacitance was 2.3 aF, estimated from the Coulomb oscillations period of 70 mV. It was possible to observe either single-island or double-island operation by adjusting the gate voltage. Clear current peaks, symmetrical in the applied bias, were observed in the Coulomb staircase. These features were associated with resonant tunnelling through quasi-bound states in the islands.

3.2.1.2 Pattern-dependant oxidation

Thermal oxidation of a nanoscale device can affect the morphology of the device in a variety of ways, some of which may be exploited to engineer ultra-small devices. Firstly, oxidation of the device after etching reduces the device dimensions, relaxing the resolution requirements for lithography to an extent. Good control over the oxidation process is essential if very small islands ~10 nm in size are required for room temperature SET operation. Secondly, the oxidation process can be strongly dependant on the shape of the device, which can be exploited to create very small islands. This PADOX process (Namatsu *et al.*, 1995; Takahashi *et al.*, 1995, 1996a, 1996b, 2002; Fujiwara *et al.*, 1998), was used to obtain SETs operating at room temperature, where the island size was only 10 nm (Takahashi *et al.*, 1995). Finally, the extent of the oxidation can be self-limited in very small structures, as increasing stress in the nanostructure prevents further oxidation. This effect is also utilized in the PADOX devices of Takahashi *et al.*.

A schematic diagram of the PADOX device of Takahashi *et al.* is shown in Fig. 3.7(a–b). These devices consist of very small nanowires, only ~10 nm wide. The wire length could be varied from 50 to 200 nm (Takahashi *et al.*, 1996a). The devices were fabricated in a SIMOX wafer with a 30 nm-thick top silicon layer (thinned from 100 nm by repeated oxidation) and a 400 nm-thick BOX layer. Electron-beam lithography, electron-cyclotron resonance (ECR) plasma oxidation and ECR plasma etching were used to define the devices. Once a nanowire had been defined in the top silicon layer of the SIMOX material, the PADOX process was used to define an ultra-small ~10 nm island in the centre of the nanowire, isolated by tunnel barriers at the ends of the nanowire. PADOX was carried out in dry oxygen ambient at 1,000°C. A top polysilicon gate was fabricated above the nanowire region to control the device current. Phosphorous ion implantation was used to dope the contact regions on either side *n*-type. The nanowire region was shielded by the top-gate and remained intrinsic. The final device could be operated using both the back- and top-gates.

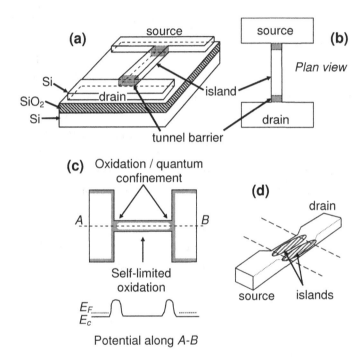

Fig. 3.7 (a) Schematic diagram of the room-temperature SET of Takahashi *et al.* (Takahashi *et al.*, 1995). An island is formed along a nanowire by tunnel barriers at the ends of a nanowire, using a PADOX process. A top-gate (not shown) controls the current. (b) Plan view of device. (c) Island formation in the centre of the nanowire, using PADOX. The potential along *A–B* is also shown. E_c is the bottom of the conduction band and E_F is the Fermi energy in the source and drain regions. (d) Twin-island vertical PADOX (V-PADOX) device (Ono *et al.*, 2000a).

The mechanism for the formation of the island and tunnel barriers by the PADOX process appears to be somewhat complicated. Initially it was believed (Takahashi *et al.*, 1995, 1996a) that during PADOX, the nanowire was oxidized from both the top and bottom, as oxygen atoms could reach the bottom of the nanowire by diffusion through the exposed BOX layer on either side. This behaviour was assumed to be most pronounced at the pattern edges (Fig. 3.7[c]). However, in the middle of the nanowire, oxidation was limited by increasing mechanical stress, ultimately preventing any further oxidation. The result was a single island in the centre of the wire, isolated by SiO_2 tunnel barriers at the contacts. However, more recently Horiguchi *et al.* (Horiguchi *et al.*, 2001) have looked at the effect of stress on the band gap of the silicon along the nanowire. They calculated the effect of both quantum confinement, associated with the reduction in the dimensions of the silicon region, and the reduction in the silicon band gap, associated with an increasing compressive stress in the middle of the nanowire due to oxidation along the length of the nanowire. The quantum confinement effect increased the energy of the bottom edge of the conduction band in the nanowire, while a compressive stress of ~20,000 atm in the centre of the nanowire decreased the energy of the bottom edge of the conduction band. The result was a potential well in the nanowire centre, isolated by tunnel barriers ~50 meV at the nanowire ends (Fig. 3.7[c]).

The island in PADOX devices is only ~10 nm, or even less, in size. As a consequence, Coulomb oscillations can occur in these devices even at room temperature. The Coulomb gap in the device of Takahashi *et al.* (Takahashi *et al.*, 1995) was ~70 meV, corresponding to a total island capacitance of only ~2 aF. The oscillation peak height increased with increasing gate voltage. This was associated with an enhancement of the carrier concentration in the contact regions, due to the formation of MOSFETs in these areas by the overlap of the top-gate (Fujiwara *et al.*, 1998).

It is possible to modify the PADOX process by simply changing the original pattern for oxidation (Fujiwara *et al.*, 1995; Namatsu *et al.*, 1996; Ono *et al.*, 2000a, 2000b). For example, Ono *et al.* obtained twin SETs rather than just a single device using a V-PADOX process, where the pattern dimensions were modulated in the vertical plane rather than in the horizontal plane (Ono *et al.*, 2000a, 2000b). Figure 3.7(d) shows a schematic diagram of the device. Here, the thickness of a somewhat wider (≥60 nm) nanowire was reduced by ~10 nm using a shallow-etched trench transverse to the nanowire (along dashed lines), creating a thinner silicon region between two thicker silicon regions. PADOX then led to oxidation of the centre of the thin silicon region and the formation of two islands along the edges of the thin silicon region, creating two SETs in parallel. The thicker silicon regions did not contain islands and formed the source and drain regions. V-PADOX was used to fabricate two SETs in a very small ~50 nm square area, illustrating the potential for very high integration levels in SET-based circuits. If only a single SET was required, V-PADOX, using a modified T-shaped pattern for the thin silicon region, could be used. Two such SETs in series have been used to fabricate a current switching device (Ono et al., 2000b).

3.2.2 SETs using MOSFET structures

A third scheme to fabricate SETs involves using patterned top-gate electrodes to define the island lithographically in the inversion layer of silicon MOSFETs (Fig. 3.8) (Matsuoka et al., 1994; Matsuoka and Kimura, 1995). This technique is an extension of the narrow channel MOSFETs of Scott-Thomas et al. (Scott-Thomas et al., 1988). Similar methods are widely used to define quantum dots in 2-DEGs in GaAs/AlGaAs heterostructure material for mesoscopic physics investigations (e.g. Kouwenhoven, 1997). In GaAs/AlGaAs 2-DEGs, the low-temperature mean free path of electrons is ~100 nm or greater, an order of magnitude longer than in a silicon inversion layer. This implies that quantum confinement and interference effects can be observed in relatively large quantum dots ~100 nm in size, albeit at very low milli-Kelvin temperatures as the charging and electron confinement energies are small. The increased complexity in measurement due to the requirement of low milli-Kelvin temperatures is counterbalanced by easier lithography requirements for the larger device sizes.

In contrast to quantum dots in GaAs/AlGaAs 2-DEGs, there is relatively limited use of similar devices in silicon inversion layers for mesoscopic physics experiments. This is a consequence of the smaller mean free path of electrons in silicon inversion layers, requiring correspondingly smaller island and device dimensions for the observation of quantum confinement effects. However, the MOS nature of these devices would raise the possibility of LSI applications. The fabrication process is also very similar to that for conventional MOSFETs.

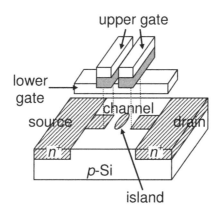

Fig. 3.8 SET in the inversion layer of a dual gate MOSFET (Matsuoka and Kimura, 1995).

Matsuoka and Kimura (Matsuoka and Kimura, 1995) have fabricated SETs in a silicon inversion layer using a dual gate device, shown schematically in Fig. 3.8. The source and drain n^+ regions in the p-Si substrate were 5 μm apart. A dual gate structure, consisting of a lower and upper gate, was placed above the region between the source and drain. Here, a 0.12 μm-wide lower gate was supported on 10 nm-thick SiO_2. This gate was used to create a narrow inversion layer between the source and drain region by applying a positive bias. The upper gate, consisting of two arms 0.1 μm apart, was supported on the lower gate by a 50 nm-thick SiO_2 layer. This gate was used to create tunnel barriers in the inversion layer by applying a negative voltage. The island area, defined by the field effect, was estimated to be 3,500 nm^2, much smaller than the geometrical size of 10,500 nm^2 defined by the dual gates. The total capacitance of the island was 28 aF, and the corresponding charging energy was 2.9 meV. This relatively small value limited single-electron effects in the device to about 4.2 K. The basic structure could also be easily changed to obtain a chain of islands forming an MTJ device, simply by increasing the number of arms of the upper gate (Matsuoka et al., 1994). However, far smaller islands would be necessary to raise the operating temperature to room temperature.

Single-electron effects have also been observed in SOI MOSFETs at low temperature (Ohata et al., 1997; Lee et al., 1998; Peters et al., 1998; Park et al., 1999,). For example, Peters et al. (Peters et al., 1998) have fabricated sub-micron SOI devices using wrapped gates. The top silicon layer of the SOI material, lightly doped p-type (10^{15} cm^{-3}), was etched to form a μm-scale wire ~100 nm thick and 500 nm wide. Wider regions, doped n-type with arsenic, formed the contact regions. The wire was then oxidized to form 25 nm-thick SiO_2 on the exposed surfaces. A polysilicon gate, wrapped around three sides of the wire defined a 200 nm-long channel. The devices channel is wider than it is long. At 1.8 K, periodic oscillations were observed in some of the devices. These oscillations could be attributed to a quantum dot defined by potential fluctuations in the channel.

It is also possible to use multiple gates to define a number of quantum dots lithographically in SOI material, following the design of Matsuoka et al. (Matsuoka et al., 1994). The devices of Lee et al. (Lee et al., 1998) and Park et al. (Park et al., 1999) used a dual gate structure, fabricated above a ~100 nm-wide wire, defined in the 80 nm-thick top silicon layer of SOI material. Multiple gates were used to define quantum dots in the inversion layer formed by a continuous gate. By adjusting the bias on the gates, single, double and triple quantum dot behaviour was observed in the electrical characteristics at 4.2 K (Park et al., 1999). Single-electron effects at 4.2 K have also been observed in edge MOSFETs, defined on the side of a ~15 nm-thick SOI layer (Ohata et al., 1997). In these devices, the island width was less than 15 nm and the length was ~100 nm.

3.2.3 Crystalline silicon nanowire SETs

Single-electron effects can be inherent to conduction in silicon nanowires defined in SOI materials, and a lithographically defined island is not essential. In these systems, the mechanism for the formation of the tunnel barriers is not very clear, and has been attributed to a variety of mechanisms. These include disorder in the doping, fluctuation in the surface depletion, quantum size effects or the formation of regions of SiO_x (Smith and Ahmed, 1997a; Ishikuro and Hiramoto, 1999). In a manner similar to etched-island SETs in SOI material, a variety of gate schemes can be used, e.g. top-gates (Ishikuro *et al.*, 1996) or side-gates (Smith and Ahmed, 1997a). With top-gates supported on SiO_2 dielectric layers, the devices are very similar to nanometre-scale SOI MOSFETs, where not only the channel length but also the channel width is of nanometre-scale. With very thin or narrow nanowires, single-electron effects can be observed at close to or even at room temperature (Ishikuro *et al.*, 1996).

Ishikuro *et al.* (Ishikuro *et al.*, 1996) have developed a well-controlled anisotropic etching process to define very narrow ~10 nm-wide wires in SOI material. The technique is shown in Fig. 3.9. A mesa was defined in the 40 nm-thick top silicon layer of the SOI material using a 10 nm-thick Si_3N_4 mask (Fig. 3.9[a]), and anisotropic wet etching in tetra-methyl-ammonium-hydroxide (TMAH). The etching process formed very smooth, sloping mesa edges along the (111) plane. The edges were then oxidized to form 30 nm-thick SiO_2 layers (Fig. 3.9[(b)]). The Si_3N_4 mask was then removed transversely across the middle of the mesa by chemical dry etching and a second anisotropic etch performed to leave two nanowires along the edges of the original mesa (Fig. 3.9[c]). The nanowires were triangular in cross-section and only ~10 nm thick and 100 nm long. The un-etched portions of the mesa at either end of the wires formed the source and drain regions. A gate oxide was then grown thermally and polysilicon top-gates deposited by low-pressure chemical vapour deposition (LPCVD) on both the wires. Ion implantation, self-aligned to the gate, was used to dope the source and drain regions *n*-type. The nanowire region remained un-doped. It is interesting to note that by simply varying the Si_3N_4 pattern, different nanowire patterns could be obtained, e.g. parallel, cross-shaped, or T-shaped nanowires (Hiramoto *et al.*, 1996).

The original device of Ishikuro *et al.* (Ishikuro *et al.*, 1996) showed sharp, pronounced, single-electron current oscillations at 4.2 K. This behaviour was attributed to both quantum confinement and single-electron charging effects in a chain of multiple quantum dots formed along the nanowire. With increasing temperature, the quantum confinement levels in the dots smeared-out first, with sharp current peaks combining to form wider periodic peaks by 77 K. The periodic peaks were attributed to single-electron charging effects only.

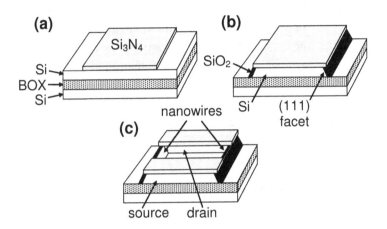

Fig. 3.9 Nanowire SET using anisotropic etching (Ishikuro et al., 1996). (a) An Si_3N_4 mask is deposited on SOI material. (b) Anisotropic etching along the (111) facets, followed by thermal oxidation of the facets. (c) Part-removal of the Si_3N_4 mask, and further anisotropic etching creates dual nanowires. A top-gate structure (not shown) can then be deposited on the nanowires.

Single-electron effects persisted in the device up to room temperature. The single-electron level spacing in the dots was estimated to be ~50 meV, greater than $k_B T$ at room temperature. The presence of strong quantum confinement effects in these devices was investigated further using very short nanowires, forming 'point-contacts' (Ishikuro and Hiramoto, 1997; Saitoh et al., 2001a). In these devices, only a single quantum dot existed, with clear 'Coulomb diamond' regions of electron stability (Fig. 3.10) in the differential conductance g_{ds} vs. V_{gs} and V_{ds} (Ishikuro and Hiramoto, 1997). The maximum Coulomb gap was ~0.2 V. However, outside the Coulomb blockade regions, negative differential conductance (NDC) regions and fine structure formed lines parallel to the edges of the Coulomb diamonds (shown schematically in Fig. 3.10). These lines were attributed to resonant tunnelling through quantum levels associated with a dot only ~6 nm in size.

The above design scheme has been used to fabricate devices with both *p*- and *n*-type source and drain regions connected to the same nanowire (Ishikuro and Hiramoto, 1999). In this novel arrangement, the observation of single-electron current oscillations in both electron and hole transport through the *same* channel suggested that the tunnel barriers were more likely to be associated with lateral quantum confinement effects or regions of SiO_x rather than surface depletion effects.

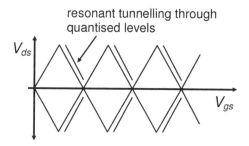

Fig. 3.10 Schematic diagram of Coulomb diamonds traced by the positions of peaks in g_{ds} vs. V_{gs} and V_{ds}. Additional lines parallel to the Coulomb diamond edges may be associated with resonant tunnelling through quantized energy levels on the island.

It is also possible to fabricate nanowire SETs, using heavily-doped nanowires. In these devices, the carrier concentration can vary from $\sim 10^{18}$ cm^{-3} to $\sim 10^{21}$ cm^{-3}. Smith and Ahmed (Smith and Ahmed, 1997a) have fabricated side-gated nanowire SETs in SIMOX SOI material with a ~50 nm-thick top silicon layer. The top silicon layer was heavily doped *n*-type using phosphorous implantation and the carrier concentration was approximately 7×10^{18} cm^{-3}. E-beam lithography and RIE was used to define the nanowire, with two in-plane side-gates on either side of the nanowire. The nanowire widths varied from ~60 to 70 nm, and the length varied from 4 to 6 μm. These rather long nanowires formed MTJs, associated with disorder along the wires. This was reflected in the complex nature of the single-electron current oscillations in the system.

A Coulomb gap could be observed in the system up to 46 K. An improved device (Smith and Ahmed, 1997b) used a nanowire doped n-type at ~10^{19} cm^{-3} with an etched width of 50 nm, thinned further using oxidation. More regular oscillations and an inverting voltage gain as high as 3.7 were observed in this device.

The tunnel barrier in these devices has been investigated experimentally by Altebaeumer and Ahmed (Altebaeumer and Ahmed, 2003), using much shorter nanowires ~80 nm in length, doped n-type at 2×10^{19} cm^{-3}. The wire widths were 35 nm after etching, reducing to only 25 nm after oxidation. A PADOX effect was expected to form tunnel barriers at the ends of the wire. However, as a function of gate voltage, the single-electron current oscillations showed a transition from a single island to a double island system. This could be attributed to fluctuation in the electrical potential along the length of the nanowire, introducing an additional tunnel barrier for part of the gate voltage range used. The potential fluctuations were attributed to changes in the dopant concentration along the wire. The effect of fluctuations in the potential on the nanowire SET characteristics, and the transition from a single dot to an MTJ, has also been investigated theoretically by Müller et al. (Müller et al., 1999, 2000) and by Evans et al. (Evans et al., 2001).

Furthermore, as the doping concentration is increased to ~10^{20} cm^{-3} or higher, the nanowires behave like metallic rather than semiconducting systems. Tilke et al. (Tilke et al., 1999) have fabricated side-gated nanowire SETs doped at 5×10^{20} cm^{-3} and observed very regular, quasi-metallic single-electron oscillations.

3.2.4 Room temperature Coulomb oscillations with large peak-to-valley ratio

A reduction in SET island size to ~5 nm or even less can lead to a dramatic improvement in the performance of a room-temperature SET. At this scale, the total island capacitance C may be ~1 aF or even smaller, with single-electron charging energies $E_c \gg k_BT$ ~26 meV at 300 K. Furthermore, the extremely small size of the island can lead to significant quantum confinement of electrons, where the energy level spacing or excitation energy E_s is also greater than k_BT at room temperature. The island then forms a quantum dot operating at room temperature.

The Coulomb oscillations in these devices can have large peak-to-valley ratios (PVRs) up to ~100 even at room temperature, much greater than the very small PVRs observed in earlier room-temperature SETs (Takahashi et al., 1995; Ishikuro and Hiramoto, 1997; Zhuang et al., 1998). Usually only a few large current peaks can be observed at room temperature. These simpler characteristics may be more advantageous for device applications.

SETs with large PVR Coulomb oscillations at room temperature have been fabricated using various means to define the islands. These include islands defined using an undulating ultra-thin SOI film (Uchida et al., 2003), ultra-small islands defined by e-beam lithography (Kitade et al., 2005) or islands defined in point-contacts (Saitoh et al., 2004). Uchida et al. (Uchida et al., 2001, 2003) have used an ultra-thin (~3 nm thick) SOI film, where surface roughness defined islands as small as ~4 nm in diameter. A combination of oxidation, followed by wet etching in an alkaline-based solution, was used to prepare an SOI film only ~3 nm thick, with undulations in the surface leading to large variations in the film thickness. The thickness variations corresponded to changes in the quantum confinement effect, with a maximum change in confinement energy of 0.25 eV. This is well above k_BT ~26 meV at 300 K. The potential in regions where the film was thinner could then be up to 0.25 eV higher than in regions where the film was thicker, leading to islands isolated by tunnel barriers even at room temperature. The lateral extent of the undulations determined the island diameter. The smallest island size in these devices was ~4 nm, with capacitance C ~0.38 aF. The room temperature PVR of the Coulomb oscillations in these devices could be as high as ~100. The charging energy was ~0.14 eV and the quantization energy was ~0.07 eV, both well above k_BT at room temperature.

Saitoh et al. (Saitoh et al., 2001b, 2004) have observed large PVR Coulomb oscillations at room temperature in SETs defined in SOI material, where an ultra-small point-contact channel defines a single island. Saitoh et al. have estimated an island size of ~4 nm, with E_c = 168 meV and E_s = 91 meV, both well above k_BT ~26 meV at 300 K

(Saitoh et al., 2001b). A PVR ~40 was observed in the Coulomb oscillations in a single-hole transistor, associated with an island only ~2 nm in size (Saitoh et al., 2004). In these devices, a large negative differential conductance peak, with a PVR of almost 12, was observed in the I_{ds}-V_{ds} characteristics, attributed to resonant tunnelling through quantized energy levels in the island.

Large PVR Coulomb oscillations have also been observed in SETs consisting of a chain of ultra-small islands, defined in heavily-doped SOI material using e-beam lithography, RIE and wet etching (Kitade et al., 2005). Here, the islands were ~20 nm in width, separated by narrower silicon regions ~10 nm in width. The island centre–centre distance was 250 nm. SETs with a single-island, and with a chain of 22 islands (forming an MTJ) were characterized. A PVR of 3.5 was observed in the single-island device, and a PVR of 77 was observed in the MTJ device.

3.2.5 Fabrication and characterization of nanowire SETs

We will now consider in detail the fabrication and electrical characterization of heavily-doped nanowire SETs in SOI material. The devices were fabricated by e-beam lithography in SIMOX SOI material. The top silicon layer of the SOI wafer, 200 nm-thick as grown, was thinned by oxidation and wet etching to 50 nm. The BOX layer was 350 nm thick. The top silicon layer was doped n-type to a doping density of 2×10^{19} cm^{-3} by implantation of Phosphorous. The devices were fabricated using 5 mm-square chips of the SOI wafer. A total of 36 devices were fabricated on each chip, organized as shown in Fig. 3.10.

The devices were fabricated within 9 large-area mesas, defined using optical lithography and RIE. Etched registration marks were defined near the mesas in the same optical mask pattern, for use in later stages of lithography. Further registration marks using silver were defined for the high-resolution e-beam lithography stages. Each mesa contained 4 SETs, defined and isolated using e-beam lithography and RIE. Each SET used a source and drain terminal connecting to the nanowire, and two gate terminals connecting to two side-gates on both sides of the nanowire (Fig. 3.11). The 4 SETs were contacted using 12 bond pads defined by optical lithography, with some sharing of the gate and drain terminals between the SETs. For example, with reference to Fig. 3.11, SET1 uses source S1, drain D1,2, gate G1 and common-gate CG1. The terminal D1,2 also forms the drain terminal for SET2, and the common-gate CG1 forms one of the side-gates for SET4.

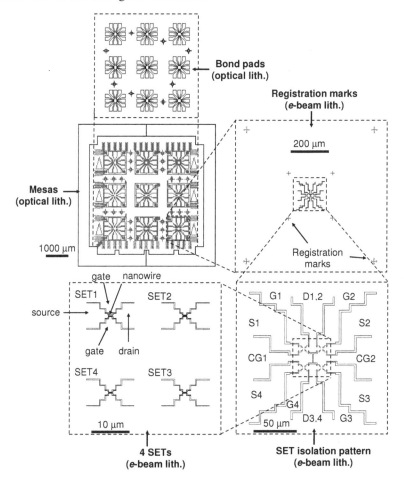

Fig. 3.11 Lithography patterns for nanowire SET fabrication.

3.2.5.1 Fabrication

The fabrication process flow is summarized in Fig. 3.12. The large-area mesas and registration marks were defined in the first fabrication stage as follows. The chips were cleaned using an ultrasonic bath rinse in acetone, followed by a similar rinse in isopropyl alcohol. Alternatively, a standard 'RCA' clean was used. After a pre-bake at 80°C to remove moisture, Shipley S1813 optical resist was spun-on at 5000 rpm and baked at 80°C for 20 minutes, giving a resist layer ~1.4 µm in thickness. Optical lithography for the mesa pattern was then carried out using a UV optical aligner, followed by development of the pattern in Shipley MF319 developer. The pattern was then transferred to the chip using RIE in a 1:1 mixture of $SiCl_4$ and CF_4. The flow rate of both gases was 20 sccm. The etching process was carried out at 300 W and 13.56 MHz for 7 minutes, long enough to etch through both the top silicon and the BOX layers. This produced very good electrical isolation between the mesas. Finally, the resist was removed using an ultrasonic rinse in acetone and isopropyl alcohol IPA.

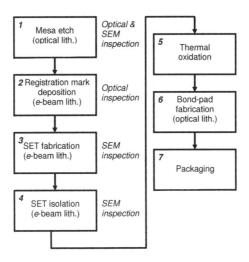

Fig. 3.12 Nanowire SET in SOI material: Fabrication process flow.

The next stage of the fabrication process defined the Ag registration marks (Fig. 3.11) for later stages of e-beam lithography. These marks were themselves fabricated using e-beam lithography, aligned to the mesas using the optical registration marks defined along with the mesa. The process was carried out in A8 PMMA 'positive' resist (1:1, 950k molecular weight, PMMA in Anisole) spun-on at 5000 rpm and baked at

180°C for a time between 1 and 20 hours. The e-beam lithography used a dose of 600 μC/cm^2 and the registration mark patterns were developed in 1:3 methyl-isobutyl-ketone (MIBK):IPA. The patterns were cross-shaped with arms ~250 nm wide. A 60 nm-thick Ag layer was then deposited by thermal evaporation, with excess metal lifted off in acetone to define the registration marks on the mesas.

Next, the SETs were patterned using e-beam lithography with a higher resolution. Again, A8 PMMA in Anisole was used, baked at 180°C for at least 1 hour. E-beam lithography at 80 kV and a beam diameter of less than 10 nm, aligned to the silver alignment marks, was used to define the nanowire and side-gate patterns with a 50 μm square field size, at a dose of ~500 μC/cm^2. The patterns were developed in 1:3 MIBK:IPA for 10 seconds in an ultrasonic bath, and defined the areas where the top silicon would be etched away down to the BOX to form the nanowire, drain, source and side-gate regions. Nanowire widths of ~50 nm and lengths from 100 to 1 μm were used. The gate to nanowire separation was ~100 nm. An RIE etch similar to that used to define the mesas, but only 40 seconds long, was used to define the SETs in the top silicon layer. Finally, the PMMA resist was removed using an ultrasonic rise in acetone and IPA.

The SETs were isolated from each other using a further stage of e-beam lithography. The patterns (Fig. 3.11) were transferred into the top silicon using A8 PMMA in Anisole at a dose of ~600 μC/cm^2 and RIE for 1 minute 15 seconds. The SETs were then inspected in an SEM to characterize the dimensions. Fig. 3.13 shows an SEM image of an SET after etching. The nanowire is ~50 nm wide and 800 nm long between the side-gates. The wire to side-gate separation is ~80 nm. The wire width increases rapidly on either side of the gated regions to avoid island formation and minimize contact resistance in these areas.

The SETs were then oxidized to passivate surface states, repair etch damage and reduce the cross-sectional area of the nanowire. Before oxidation, the Ag registration marks were etched away in a 1:1:4 solution of H_2O_2, 35% ammonia and methanol in an ultrasonic bath, followed by further cleaning in acetone and IPA, and in an O_2 plasma etcher. Oxidation was carried out at 1000°C for 15 minutes, reducing the thickness and width of the silicon in the nanowire from 50 nm to ~30 nm.

Finally, 400 nm-thick bond-pads were defined on the mesa, using optical lithography, Al evaporation and lift-off of the excess metal. A short oxide-etch in SILOX SiO₂ etchant was carried out before the metal evaporation to obtain a good Ohmic contact. The chip was then diced into smaller pieces, each containing a single mesa, and packaged in a 20 pin chip carrier, with 4 SETs per carrier.

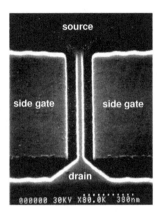

Fig. 3.13 Scanning electron micrograph of a nanowire SET in SOI material, before thermal oxidation.

3.2.5.2 Electrical characterization

Figure 3.14 shows the electrical characteristics at 4.2 K, 77 K and 300 K for a 40 nm-wide (as etched) and 800 nm-long silicon nanowire SET. The silicon core of the nanowire, after oxidation, was ~20 nm × 30 nm. The SET was characterized electrically using an Agilent HP4156 parameter analyser.

Figure 3.14(a) shows the I_{ds}-V_{gs} characteristics of the device at 4.2 K, 77 K and 300 K, at V_{ds} = 20 mV. A positive gate voltage is necessary for conduction through the device. This voltage increases from ~2 V at 300 K to ~10 V at 77 K. As the silicon core of the nanowire is only ~20 nm wide, surface depletion effects and any fluctuations in the doping potential are likely to deplete the nanowire completely of electrons. This raises the bottom of the conduction band E_c in the nanowire relative to the Fermi energy E_F in the drain and source regions. A positive threshold voltage for conduction is then observed. In addition, any variation in the oxidized width, e.g. by PADOX and lateral quantum confinement at the ends of the nanowire (Horiguchi *et al.*, 2001), can lead to potential barriers at the nanowire ends (Altebaeumer and Ahmed, 2003).

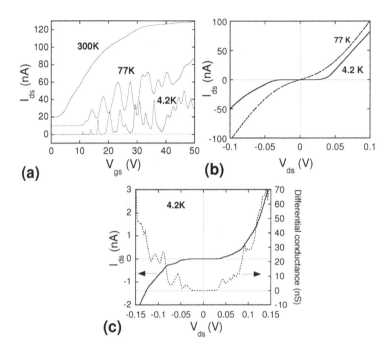

Fig. 3.14 *I-V* characteristics of a nanowire SET in SOI material. (a) I_{ds}-V_{gs} characteristics of the device at 4.2 K, 77 K and 300 K, at V_{ds} = 20 mV. The curves are offset by 10 nA per temperature step for clarity. (b) I_{ds}-V_{ds} characteristics of the device at 4.2 K and at 77 K. (c) Coulomb staircase from a second device at 4.2 K, observed in the I_{ds}-V_{ds} (solid line) characteristics, and the differential conductance, $\partial I_{ds}/\partial V_{ds}$-$V_{ds}$, (dotted line) characteristics.

The I_{ds}-V_{gs} characteristics of the device at 4.2 K show periodic single-electron current oscillations. The first current peak is observed at V_{gs} = 11 V. A single oscillation period of ΔV_{gs} = 3 V is observed up to V_{gs} = 24 V, corresponding to the fifth current peak. The corresponding gate-island capacitance is very small, $C_g = e/\Delta V_{gs}$ = 0.053 aF. The single oscillation period in this range implies the existence of a single charging island along the wire. Above V_{gs} = 24 V, the oscillations become more complex, with the introduction of additional peaks with a smaller oscillation period. Above V_{gs} = 40 V, it becomes difficult to observe the oscillations observed for V_{gs} < 24 V. This behaviour may be associated with the formation of an MTJ (Altebaeumer and Ahmed, 2003). There is also a gradual increase in the average current, associated with an increase in the carrier concentration in the nanowire due to the field effect of the gate (Müller et al., 1999, 2000). Strong current oscillations are also observed at 77 K, though the peaks are thermally broadened and there is a rise in the thermally activated current. The peak position and oscillation period also vary slightly, due to changes in C_g. At 300 K, the oscillations are almost completely thermally smeared out. However, very slight changes in the slope of the I_{ds}-V_{gs} curve can still be observed, corresponding to traces of the oscillations at lower temperatures.

Figure 3.14(b) shows the I_{ds}-V_{ds} characteristics of the device at 4.2 K and at 77 K. A symmetrical Coulomb gap V_{cg} = 70 mV is observed at 4.2 K and V_{gs} = 17 V. The gap changes into a non-linearity at 77 K due to an increased thermally activated current. A Coulomb staircase is not observed, suggesting that the tunnel barriers connecting the island to the leads are very similar. The total island capacitance corresponding to this is $C_t = 2e/V_{cg}$ = 4.6 aF, where we have assumed equal tunnel capacitors (see Chapter 2). The charging energy of the island $E_c = e^2/2C_t$ = 17 meV. This is very close to k_BT = 25 meV at 300 K and explains why slight traces of single-electron charging effects persist even at 300 K in the device. Depending on the nature of the tunnel barriers in these devices, a Coulomb staircase can also be observed. Figure 3.14(c) shows the characteristics from a different device, where a clear Coulomb staircase is observed. Here, the Coulomb gap is ~50 meV.

Very strong single-electron effects can be observed in nanowire SETs with reduced width. In doped nanowire devices, an increase in the doping density is necessary to allow observation at low gate voltages. Figure 3.15 shows the Coulomb staircase characteristics at 4.2 K in a 100 nm-long side-gated nanowire SET, where the cross-section was only ~10 nm after oxidation. In this device, the top silicon layer of the SOI material was very heavily-doped n-type, at a concentration of ~1 × 10^{20} cm^{-3}. Figure 3.15 shows the I_{ds}-V_{ds} characteristics of the device as the side-gate voltage V_{gs} is varied from 0 V to –10 V. The curves are offset 10 nA per gate step of 0.1 V for clarity. Because of the heavier doping, this device conducts at V_{gs} = 0 V. As the gate voltage is varied from 0 V to –9 V, six Coulomb diamonds are observed. For the first four Coulomb diamonds, from V_{gs} = 0 V to V_{gs} = –5.5 V, a Coulomb gap V_{cg} ≈ 100 mV is observed. The gap can be modulated to zero at specific values of the gate bias. This part of the characteristic can be attributed to a single charging island, of capacitance $C_t = 2e/V_{cg}$ = 3.2 aF and charging energy $E_c = e^2/2C_t$ = 25 meV. The period of the Coulomb diamonds in gate voltage, corresponding to the single-electron current oscillation period, is ΔV_{gs} ≈ 1.3 V. The associated gate-island capacitance $C_g = e/\Delta V_{gs}$ = 0.12 aF.

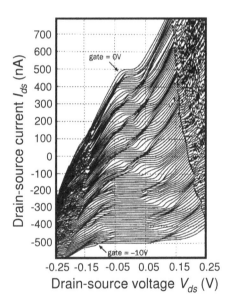

Fig. 3.15 Electrical characteristics of a 100 nm-long nanowire SET in SOI material at 4.2 K. The island is ~10 nm in size. (Smith, 1998).

As the gate voltage decreases from V_{gs} = –5.5 V to V_{gs} = –10 V, two complete and one almost complete Coulomb diamond can be observed. Here, the diamonds cannot be fully modulated by the gate voltage, this behaviour becoming more prominent at more negative gate voltages.

This may be attributed to potential barriers in series, e.g. due to the formation of an MTJ. Furthermore, additional lines parallel to the edges of the Coulomb diamonds appear, formed by small steps in the current. This fine structure is a signature of quantum confinement effects in the island, i.e. the formation of a quantum dot (Ishikuro and Hiramoto, 1997; Saitoh *et al.*, 2001a). The edge of the Coulomb diamond corresponds to resonance of the ground state of the quantum dot with the Fermi energy of the source, and the fine structure in parallel at higher values of V_{ds} corresponds to resonance of the first excited state of the quantum dot with the Fermi energy of the source.

3.3 Single-Electron Transistors in Nanocrystalline Silicon

Nanocrystalline silicon (nc-Si) materials (Grom *et al.*, 2000; Littau *et al.*, 1993; Wilson *et al.*, 1993; Oda *et al.*, 1995; Otobe *et al.*, 1995; Yano *et al.*, 1995, 1999; Nakajima *et al.*, 1996; Tiwari *et al.*, 1996a; Kamiya *et al.*, 1999, 2001; Nakahata *et al.*, 2000) consist of crystalline silicon grains ~10 nm in size, separated by oxide or amorphous regions. These materials provide an alternative means to fabricate single-electron devices, where the grains form the charging islands and the oxide or amorphous regions form the tunnel barriers. The islands and tunnel barriers can then be defined using growth techniques rather than high-resolution lithography, with the possibility of better control over their dimensions.

We will use the term nc-Si to refer to two types of materials. The first type consists of continuous thin films of nanocrystalline Si, very similar to those used for thin film transistor and display applications (Fig 3.16[a]) (Kamiya *et al.*, 1999, 2001). Here, nanoscale crystalline silicon grains are separated by grain boundaries (GBs) consisting of thin amorphous or silicon oxide tissues. The second type of material consists of discreet silicon nanocrystals, either deposited in layers to form a strongly non-homogeneous film (Fig. 3.16[b]), or used individually in the device in combination with high-resolution lithography. Here, the silicon nanocrystals may be surrounded by well-defined oxide shells (Oda *et al.*, 1995; Otobe *et al.*, 1995), or may be embedded in an insulating matrix (Tiwari et al., 1996a). For both types of materials, if the silicon nanocrystals are ~10 nm or less in size, and the tunnel barriers are ~100 meV or higher, then the single-electron charging energy and tunnel resistances can be large enough for the fabrication of room-temperature SETs (Yano et al., 1995; Tan et al., 2003). In nanocrystals of this scale, quantum confinement of electrons may also occur, raising the possibility of the formation of silicon quantum dots (Kouwenhoven et al., 1997). Room temperature SETs may also be more easily realizable using nc-Si rather than by high-resolution lithography. It may be possible to control the size and shape of the silicon grains in nc-Si with a precision greater than is possible with high-resolution lithographic techniques, by carefully tailoring the material growth process (Oda et al., 1995; Otobe et al., 1995; Kamiya et al., 1999, 2001). This would help to improve reproducibility between the electrical characteristics of large numbers of devices, necessary for LSI circuit applications.

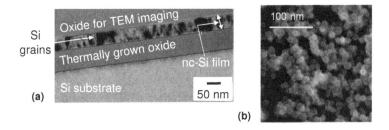

Fig. 3.16 (a) Transmission electron micrograph of a nanocrystalline Si thin film, prepared by LPCVD. The film is ~50 nm thick and doped n-type at ~10^{19}/cm^3. (b) Scanning electron micrograph of Si nanocrystals, 8 nm ± 1 nm in diameter, created by plasma decomposition of SiH$_4$ (Oda et al., 1995). The nanocrystals are deposited as a thin film from a plasma cell onto a Si substrate.

We can estimate the order-of-magnitude of the charging energy for a spherical silicon nanocrystal using the self-capacitance of a sphere. For a spherical nanocrystal of diameter $d = 10$ nm, embedded in SiO$_2$, the self-capacitance $C = \pi\varepsilon d = 1$ aF. This implies that the single-electron charging energy, $E_c = e^2/2C \sim 74$ meV, greater than $k_B T = 25$ meV at room temperature ($T = 300$ K). Furthermore, quantum confinement effects may also occur at or near room temperature, such that the nanocrystal behaves as a quantum dot even at room temperature.

Assuming that the nanocrystal forms a spherical potential well of width d = 10 nm with vertical potential barriers, electrons occupy discrete energy levels within the well, with an energy level spacing $\Delta E \sim \pi^2 \hbar^2/(2m[d/2]^2)$ = 0.04 eV, greater than the thermal energy at 300 K. This is much larger than the energy level spacing in GaAs/AlGaAs 2-DEG quantum dots, which tend to be much larger in size. Quantum confinement effects may then influence the electrical characteristics of a nanocrystal single-electron device, even at room temperature.

3.3.1 Conduction in continuous nanocrystalline silicon films

We first discuss the electronic conduction mechanism in doped, physically continuous polycrystalline or nc-Si thin films without the influence of single-electron charging effects. We then extend this picture to include single-electron and quantum confinement effects associated with nanoscale grain sizes.

Conduction through an nc-Si film is strongly affected by potential barriers at the GBs, associated with the large density of trapping states caused by defects at the GBs. These states trap free carriers from the grains, reducing the carrier density within the grain. The space charge distribution near the GB creates an electric field, causing a 'Schottky-like' potential barrier at the GBs (Kamins, 1971; Seto, 1975; Baccarani et al., 1978; Levinson et al., 1982). The height and width of the potential barrier is a function of the doping concentration in the grains. In addition to carrier trapping at the GBs, any segregation of dopant atoms at the GBs reduces the effective carrier density in the grains even further (Cower and Sedgwick, 1972; Baccarani et al., 1978; Fripp, 1975).

Consider a one-dimensional chain of n-type nc-Si grains (Fig. 3.17[a]), where the GB thickness is small relative to the grain size 'D'. We assume a uniform donor concentration N_D (per unit volume) in the grain, and GB traps with a density N_t (per unit area) at an energy E_t with respect to the intrinsic Fermi level. We note that in large grained polycrystalline silicon films, N_t is often $\sim 10^{11}$–10^{12} /cm^2 (Kamins et al., 1980; Levinson et al., 1982).

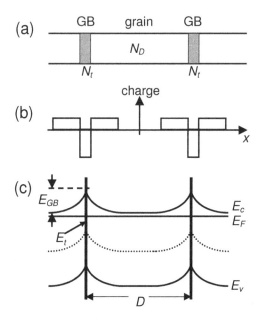

Fig. 3.17 Energy band diagram across a one-dimensional chain of n-type nanocrystalline Si grains, with trapping states at the GBs. (a) Schematic diagram, with a grain isolated by GBs. The trap density is N_t (per unit area). (b) Charge distribution across the grain and GBs. (c) Energy bands across the grain and GBs.

Electrons trapped at the GBs leave ionized donors in the grains (Fig. 3.17[b]). For small N_D, all the electrons contributed by the dopants are trapped in the GBs and the grain is fully depleted. The trapped charge and the ionized dopants generate an electric field extending from the GB into the grains, leading to a double Schottky-like potential barrier of height E_{GB} (Fig. 3.17[c]). As N_D increases, more charge is trapped at the GB, increasing the electric field and potential barrier height until at $N_D = N_D^* \approx N_t/D$, the conduction band in the centre of the grain lies near the Fermi energy E_F. Free carriers can now exist in the grain and E_{GB} is at its maximum value. Any further increase in N_D reduces E_{GB}. In the above discussion, we have assumed that N_t is high enough such that all the traps are not filled if N_D is increased. In addition, in a real nc-Si film, the grain size, GB trap density and local doping concentration is likely to vary from grain to grain, leading to a distribution of GB barrier heights and widths across the film (Tringe and Plummer, 2000).

Electron transport across the GBs at room temperature, and at moderately low temperature, can occur by thermionic emission. With this mechanism, the temperature dependence of the conductance, plotted as an Arrhenius plot, $\ln(G)$ vs. $1/T$, will be linear. However, the conduction mechanism may be assisted by tunnelling via defect states within the barrier, e.g. by empty states at the GB, or by tunnelling across the entire barrier if the barrier width is small. As the temperature is reduced, the thermionic emission current falls and tunnelling effects begin to dominate the conduction process, leading to a largely temperature-independent section of the Arrhenius plot. This is shown schematically in Fig. 3.18. The slope of the temperature-dependent section of the plot can be used to extract the activation energy, which is a measure of the barrier height. At low temperatures, a variable range hopping transport mechanism may also contribute to conduction (Shklovskii and Efros, 1984; Dong *et al.*, 2004; Rafiq *et al.*, 2006). In addition, any variation in the barrier heights and widths across the film can result in a network of percolation paths for current flow across the film (Shklovskii and Efros, 1984; Tringe and Plummer, 2000), where low resistance paths through GBs with low potential barriers dominate the conduction process.

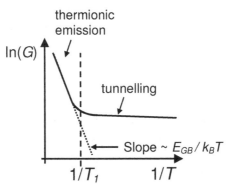

Fig. 3.18 Arrhenius plot of the nanocrystalline Si film conductance G vs. inverse temperature $1/T$, shown schematically. Thermionic emission of electrons over the grain boundaries dominates at $T > T_1$. At $T < T_1$, electron tunnelling through the grain-boundary potential barrier dominates.

At this stage, we have ignored single-electron charging effects in the silicon grains. A thermionic emission model is sufficient if the nc-Si grains are large enough, or if the temperature is high enough such that $k_BT > E_c = e^2/2C$. The thermionic emission model will also be valid if the GB barrier height and width is small enough such that the associated tunnel resistance R_{GB} is comparable to, or smaller than, the quantum resistance $R_K = h/e^2 \sim 25.9$ kΩ, i.e. electrons are delocalized across the grains.

We will now consider the effect of the nanoscale grain size in nc-Si films. Typically, single-electron nc-Si devices have grain sizes from ~50 nm to <10 nm. As the grain size is reduced to the nanometre scale, the local or 'microscopic' properties of the GBs (Furuta et al., 2001, 2002), single-electron charging effects and quantum-confinement effects all begin to affect the electron transport mechanism. We have seen that the nc-Si film may be regarded as an array of nanoscale conducting grains, isolated from each other by potential barriers at the GBs. We have also seen that, at cryogenic temperatures, electron transport can occur by tunnelling through the GB potential barriers. The nc-Si film under these conditions can then be considered to form an array of nanoscale tunnel capacitors which can show single-electron charging effects. If the grains are ~10 nm in size, then it is possible for the capacitance $C \sim 1$ aF, such that $E_c > k_BT \sim 26$ meV at room temperature.

However, for the observation of room temperature single-electron charging, it is also necessary for the GB barrier height to be considerably greater than the thermal energy k_BT, and the GB tunnelling resistance $R_{GB} > R_Q$, so that electrons can be quasi-localized on the grains at room temperature. Coulomb blockade then occurs in the I_{ds}-V_{ds} characteristics across the grain, and current flows only if V_{ds} can overcome E_c. Discrete electron energy levels on the grain can increase the threshold voltage for current even further. A parabolic potential well can describe the potential across the grain, from an extension of the simple polycrystalline silicon model discussed earlier. The grain would then behave as a silicon quantum dot and the I-V characteristics across the system would show a combination of single-electron and resonant tunnelling effects.

While single-electron and quantum-confinement effects are clearly significant in individual grains in an nc-Si film, in a large-area film, variation in the grain size and in the tunnel barriers at the GB may smear-out these effects. Furthermore, percolation conduction of electrons can occur through the lowest resistance transport paths (Furuta *et al.*, 2001, 2002). This would tend to bypass the higher resistance paths associated with any grains forming quantum dots, and prevent the observation of single-electron effects. Therefore, it is necessary in most demonstrations of single-electron charging and quantum dot devices to reduce the number of current paths by defining nanowires, or by defining 'point-contacts' (i.e. a short nanowire where the length ~ width). Depending on the geometry, one or more than one grain can contribute to single-electron charging effects, leading to single-island or MTJ devices, respectively.

3.3.2 Nanocrystalline silicon nanowire SETs

Nanowire SETs fabricated in continuous nc-Si films are analogues of nanowire SETs fabricated in crystalline SOI material (see Section 3.2.3 for details). Nanowire SETs in crystalline silicon are ~50 nm or less in width, defined by trench isolation in the top silicon layer of SOI material. The layer is often heavily doped and ~50 nm or less in thickness. The island and tunnel barriers along the nanowire may be defined by disorder associated with the doping and surface states, oxidation or unequal quantum confinement effects, resulting in an MTJ system. Alternatively, the potential barriers may be defined in a controlled manner by self-limiting oxidation, unequal quantum confinement, stress-induced changes in the band gap, or by patterning notches along the wire. The nanowire current can be gated using a variety of techniques, e.g. trench-isolated side-gates, deposited polycrystalline silicon or metal top-gates, or using the back-gate formed by the substrate of the SOI material.

Nanocrystalline silicon nanowire SETs (Fig. 3.19), very similar in geometry to their crystalline silicon counterparts, can be defined in polycrystalline or nc-Si films ~50 nm or less in thickness. The different variations of gate structures for SOI nanowires can also be applied here. However, the presence of GBs intrinsically creates tunnel barriers along nc-Si nanowires, isolating charging islands at the grains, and doping or surface disorder effects are subsidiary to this. As a secondary effect, doping and/or surface disorder can affect the GB potential barrier shape, by altering the local space charge distribution.

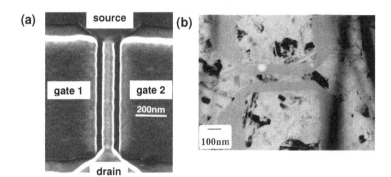

Fig. 3.19 (a) Scanning electron micrograph of a nanowire SET in polycrystalline Si. The nanowire is 1 μm long and 40 nm wide and is oxidized, with side-gates. (b) Transmission electron micrograph of a similar SET, with a nanowire 400 nm long and 50 nm wide. The polycrystalline Si grains are ~20–100 nm in size.

We now consider lateral, side-gated nanowire SETs, fabricated in solid-phase crystallized (SPC) polycrystalline silicon films deposited on SiO_2 grown on silicon substrates (Irvine *et al.*, 1998; Tan *et al.*, 2001a). Figure 3.19(a) shows an SEM image of a typical device. The polycrystalline silicon material was prepared using a standard LSI process, as follows: A 50 nm-thick amorphous silicon film was deposited first at 550°C by PECVD, onto a 10 nm- or 40 nm-thick SiO_2 layer, grown on a crystalline silicon substrate (*p*-doped, at 5×10^{14} cm^{-3}). Phosphorous ion-implantation was used to heavily dope the film *n*-type to 5×10^{19} cm^{-3}. The film was then crystallized into polycrystalline silicon using thermal annealing at 850°C for 30 minutes. Transmission electron microscope (TEM) analysis indicated that the grains varied from ~5 to 50 nm in size, and the average grain size was ~20 nm. Side-gated nanowires of various dimensions could be defined in the film, using e-beam lithography and RIE in $SiCl_4/CF_4$ plasma. Nanowires with argon annealing or oxidation treatments were also fabricated. Figure 3.19(a) shows an SEM image of this type of device after oxidation, where the nanowire is 1 μm long and 40 nm wide. The oxidation process reduced the cross-sectional area of the nanowire by ~10 nm and passivated surface states. The annealing process reduced the defect state density at the GBs and at the Si/SiO$_2$ interface, and increased the grain size. Figure 3.19(b) shows a transmission electron micrograph of a similar SET, where the nanowire is 400 nm long and 50 nm wide. The polycrystalline silicon grains are clearly visible, and are ~20–100 nm in size.

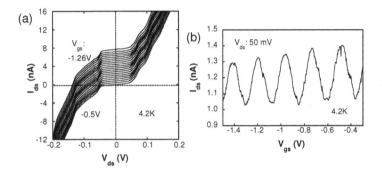

Fig. 3.20 *I-V* characteristics of a nanocrystalline silicon nanowire SET at 4.2 K. (a) I_{ds}-V_{ds} characteristics. The curves are offset 4 nA/40mV gate step for clarity. (b) I_{ds}-V_{gs} characteristics.

The drain source I_{ds}-V_{ds} characteristics at 4.2 K, from an oxidized nanowire with a pre-oxidized width of 50 nm and length of 1.5 µm, are shown in Fig. 3.20(a) (Irvine *et al.*, 1998; Tan *et al.*, 2001a). A Coulomb staircase is seen in the characteristics. Single-electron current oscillations are observed in the I_{ds}-V_{gs} characteristics (Fig. 3.20[b]), corresponding to the addition of single electrons to a dominant charging grain. The single-electron current oscillations in other devices can be complex, due to the multiple periods associated with an MTJ and due to changes in the gate capacitance with voltage. This device had a rather low maximum operating temperature of ~15 K, due to the large grain size and the low height of the GB tunnel barriers.

The single-electron characteristics in these devices are dependent on the nanowire dimensions (Tan *et al.*, 2001a). The oscillation periods increase when the nanowire length is increased from 500 nm to 1.5 µm, and decrease when the nanowire width is increased from 50 to 60 nm. Wider wires show only Ohmic conduction. This is because the longer the nanowire, the higher the probability of smaller grains existing along the nanowire, with smaller gate capacitances and larger observed oscillation periods. The decrease in the oscillation period with increasing width implies an increase in the island capacitance. This is because the gate-island capacitance has a component associated with the part of the electric field between the gate and island which passes through the buried oxide. Here, the electric field lies below the plane of the side-gates and the nanowire.

The effect of the grain size in nc-Si SETs can be observed directly in the single-electron characteristics, as this is proportional to the island capacitance and therefore determines the Coulomb gap and the current oscillation period. However, the GB tunnel barrier must be investigated by other means. Arrhenius plots of the SET conductance as a function of temperature (Fig. 3.18) can be used to extract the thermal activation energy associated with the barrier height. This technique has been used to investigate the effect of restricting the multiple current paths in nc-Si nanowire SETs, by varying the dimensions of the nanowire.

Furuta *et al.* (Furuta *et al.*, 2001, 2002) have investigated the electrical properties of a single GB at the microscopic scale, using nanowires defined by e-beam lithography in 50 nm-thick polycrystalline silicon films with grains 20–150 nm in size. Furuta *et al.* fabricated nanowires of varying width and length, from 30 to 50 nm, and measured the distribution of the potential barrier height using Arrhenius plots. They observed that any local variation in the potential barrier height of a GB provided a low resistance path for current transport across the GB. If the nanowire width was increased, a lower barrier height was measured because of the increased likelihood of a lower section of the GB barrier across the nanowire. If the nanowire length was increased, a higher barrier height was observed because more than one GB could lie in series, the GB with the highest barrier dominating the measurements.

We observe that the barrier height can vary even at various points along a single GB. This has significant implications for the fabrication of nanometre-scale electronic devices in nc-Si films. Disorder in the GB potential barrier can lead to considerable variation in the *I-V* characteristics of different devices, and a means of control over the GBs may be essential for the practical fabrication of circuits using nc-Si devices. Careful optimization of the polycrystalline silicon film growth process or post-deposition treatment of the film (see Section 3.3.4) may help to control the composition of the GBs.

3.3.3 Point-contact nc-Si SET: Room temperature operation

The 'point-contact' nc-Si SET uses a short nc-Si nanowire, where both the length and width are reduced to ~50 nm or less (Fig. 3.21[a]). At most, only a few grains can exist within the active area of the device, improving the electrical characteristics. Reducing the grain size in a point-contact nc-Si SET to ~10 nm can increase the maximum operating of the device to room temperature.

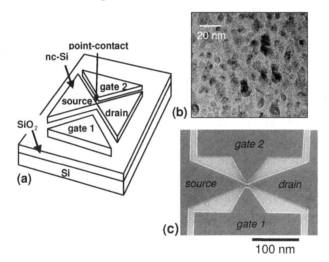

Fig. 3.21 (a) Point-contact nanocrystalline Si SET with side-gates. (a) Schematic diagram of the SET. (b) Transmission electron micrograph of the nanocrystalline Si film. (c) Scanning electron micrograph of the SET.

Nanocrystalline silicon films with grains of this scale, for SET fabrication, have been prepared using a variety of methods. These include very high frequency (VHF) PECVD films of thickness ~20 nm and grain size <10 nm (Kamiya et al., 2001), and LPCVD films of thickness ~40 nm and grain size from ~10 to 30 nm (Khalafalla et al., 2003). It is also possible to prepare ultra-thin (<10 nm) strongly granular and non-uniform films by crystallization of thin amorphous silicon layers (Yano *et al.*, 1995). We discuss SETs fabricated in strongly granular films in Section 3.3.5. In this section, we discuss the fabrication and characterization of point-contact SETs in VHF PECVD nc-Si films, and the improvement of the operating temperature of these SETs to room temperature by selective oxidation of the GBs.

Point-contact SETs operating up to 60 K have been fabricated in ~30 nm-thick nc-Si, with grains <10 nm in size (Tan *et al.*, 2001b). The films were deposited by VHF PECVD using a SiF_4:H_2:SiH_4 gas mixture, onto a 150 nm-thick silicon oxide layer grown thermally on *n*-type crystalline silicon. The carrier concentration and electron mobility, measured at room temperature by Hall measurements, were 3×10^{20}/cm^3 and 1.8 cm^2/Vs, respectively. Figure 3.21(b) shows a TEM image of the film, where uniformly distributed crystalline silicon grains can be seen. The grain size ranges from ~4–8 nm and the GBs consist of amorphous silicon tissues ~1 nm thick.

While the grain size was well below 10 nm, the maximum operating temperature of the SETs was only 60 K, the reason for which is discussed below. The crystalline volume fraction, determined by Raman spectroscopy, was 70%. Point-contact SETs were defined in these films using e-beam lithography in PMMA resist, and RIE in a mixture of $SiCl_4$ and CF_4 gases, in a manner similar to the nanowire SETs discussed in the previous section. The point-contact width was ~20 nm and two side-gates could be used to control the device characteristics. Figure 3.21(c) shows a scanning electron micrograph of a device. These devices show Coulomb gaps ~40 mV which persisted up to 60 K. The single-electron current oscillations showed a main oscillation with a period of 500 mV. Finer superimposed oscillations were also observed, attributed to the formation of an MTJ.

The operating temperature in these devices was limited not by the grain size but by the tunnel resistance and height of the GB potential barriers. The low operating temperature, even though the grain size was small enough, and the charging energy large enough for high temperature operation, could be associated with a low barrier height. The barrier height was estimated using Arrhenius plots of the device conductance (Fig. 3.22). Above a transition temperature T_1 ~60 K, the conduction mechanism could be attributed to thermionic emission across a distribution of potential barrier heights with various activation energies. The maximum gradient obtained in this region corresponded to an activation energy E_A ~40 meV, which could be associated with the maximum height of the amorphous silicon GB tunnel barriers. This value was not high enough, relative to $k_B T$ ~25 meV at room temperature, to confine electrons on the grains at room temperature.

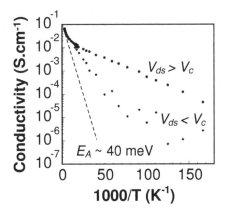

Fig. 3.22 Arrhenius plot of conductivity vs. inverse temperature in a nanocrystalline silicon point-contact SET, for V_{ds} = 50 mV > V_c, and at zero V_{ds}. Here V_c ~ 10 mV.

Selective oxidation of the amorphous silicon GBs into SiO_x can raise the operating temperature of these devices to room temperature (Durrani et al., 2002, 2003; Tan et al., 2003). Point-contact devices with dimensions of 20 nm × 20 nm × 20 nm have been prepared in slightly thinner (~20 nm) VHF PECVD nc-Si films, again with grain sizes from ~4 to 8 nm. However, after defining the SETs, a low-temperature oxidation and high-temperature annealing process was used to oxidize the GBs selectively and improve the tunnel barrier height. A relatively low oxidation temperature of 750°C for 1 hour was used, in order to take advantage of the higher rate of diffusion of oxygen atoms into the GBs at these temperatures than in the crystalline silicon grains. The devices were then annealed at 1000°C for 15 minutes, which further improved the tunnel barrier height. Microscopy of the SET before and after the thermal processing did not show significant change in the grain shape and size, due to the encapsulation of the grains by SiO_x.

The selective oxidation of the GBs provides a method to engineer GB tunnel barriers with increased potential energy, high enough to observe room temperature single-electron effects (Tan et al., 2003). The I_{ds}-V_{gs} characteristics of a selectively-oxidized device, at temperatures from 23 to 300 K, are shown in Fig. 3.23(a). Single-electron current oscillations with a single oscillation period of 3 V are seen, which can be associated with a single dominant charging island. The oscillations persist up to 300 K with an unchanged period. However, there is a fall in the peak-to-valley ratio as the temperature increases, due to a thermally activated increase in the tunnelling probability. Figure 3.23(b) shows the device I_{ds}-V_{ds} characteristics at 300 K, where a non-linear region corresponding to the Coulomb gap can be observed.

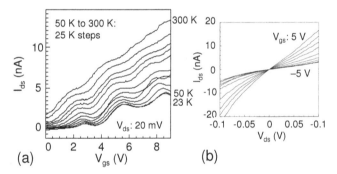

Fig. 3.23 *I-V* characteristics of a nanocrystalline Si point-contact SET with oxidized grain boundaries. (a) Temperature dependence of I_{ds}-V_{gs} single-electron oscillations, from 50 to 300 K. The oscillations persist at 300 K. (b) I_{ds}-V_{ds} characteristics at 300 K. A non-linearity corresponding to the Coulomb gap is observed at 300 K, at $V_{gs} \leq 0$ V.

The room temperature operation of these devices could be attributed to the formation of SiO_x at the GBs, leading to an increase in the tunnel barrier height and better confinement of electrons on the grains. The tunnel barrier height, measured using Arrhenius plots of the device conductance, was ~170 meV. This is approximately seven times higher than k_BT at room temperature and considerably larger than the maximum barrier height of ~40 meV for similar devices in as-deposited nc-Si films.

The oxygen incorporation in the oxidized and annealed nc-Si film was investigated using secondary-ion mass spectroscopy (SIMS) to measure the oxygen depth profile (Tan *et al.*, 2003). It was seen that in a 30 nm wide point-contact, SiO_x with $x = 0.67$ was formed. Greater amounts of oxygen could be incorporated into the GBs in smaller point-contacts, due to diffusion from the side-walls of the point-contact. It was then possible for single-electron charging to occur on grains at the point-contact centre, even at room temperature.

In the preceding discussion, we have concentrated only on single-electron effects in nc-Si SETs. However, quantum-confinement effects may also occur in the nc-Si grains. The existence of discrete energy levels within silicon nanocrystals has been inferred from observations of light-emission from the nanocrystals (Brus *et al.*, 1995; Kanemitsu, 1995; Kanemitsu *et al.* 1997). In an nc-Si SET, discrete energy levels would lead to resonant tunnelling peaks in the gate dependence of the drain-source current, where the peak separation corresponds to a sum of the single-electron and quantum-confinement energy level separation and the peak height corresponds to the coupling of the electron wavefunction associated with each energy level to the contacts. Natori *et al.* (Natori *et al.*, 2000) have theoretically investigated this behaviour for silicon dots. Interactions between two or more quantum dots are also possible, and both electrostatic and electron wavefunction coupling effects can be observed. These effects will be discussed in more detail in Section 3.3.7.

Vertical transport SETs have also been demonstrated in polycrystalline Si and nc-Si. Single-electron effects have been observed at 4.2 K in 45–100 nm diameter pillars fabricated in a material consisting of layers of polycrystalline silicon and Si_3N_4 (Pooley *et al.*, 1999). In this device, the Si_3N_4 layers form the tunnel barriers and the polycrystalline silicon layers form the charging island. The lateral dimensions of the charging islands are defined not by the grain size but by the pillar side-walls, i.e. lithographically. While the device does not show single-electron effects at room temperature, it can still be used as a vertical-transport switching device at room temperature.

A vertical transport device may have considerable advantages in device integration, e.g. such a device can be stacked on top of the gate of a MOSFET to form a random access memory gain-cell, where the number of stored electrons is only ~1,000 (Mizuta *et al.*, 2001).

3.3.4 'Grain-boundary' engineering

The previous sections have discussed the significance of the GB potential barrier in SETs in continuous nc-Si films. Control of the height of this barrier is crucial to the confinement of electrons on the grains at higher temperatures, and the fabrication of room temperature SETs. By contrast, the reduction of the GB potential barrier is important in reducing the film resistivity, and improving the effective carrier mobility in the nc-Si. Different 'GB engineering' processes can be used to address these requirements.

The effect of oxidation and annealing on the electrical properties and the structure of the GBs in heavily doped SPC polycrystalline silicon has been characterized in detail using bulk films, and using 30 nm-wide nanowires (Kamiya *et al.*, 2002). Oxidation at 650–750°C was seen to oxidize the GBs selectively, and subsequent annealing at 1,000°C was seen to increase the associated potential barrier height and resistance. These observations were explained by structural changes in the Si–O network at the GBs, and the competition between surface oxygen diffusion and oxidation from the GBs in the crystalline grains. This work suggested that a combination of oxidation and annealing provided a method for better control of the GB potential barrier height and width in polycrystalline silicon and nc-Si thin films.

In contrast, hot H_2O-vapour annealing reduced the GB barrier height (Kamiya *et al.*, 2003). Experiments on nanowire devices fabricated in LPCVD polycrystalline silicon thin films showed that hot H_2O-vapour annealing reduced the GB dangling bonds and the corresponding potential barrier height. In addition, the process narrowed the distribution of the barrier height value across different devices significantly. These effects could be attributed to oxidation in the vicinity of the film surface, and hydrogenation in the deeper regions of the film. The results suggested that H_2O annealing could improve the carrier transport properties by opening up shorter percolation paths, and by increasing the effective carrier mobility and density.

3.3.5 SETs using discrete silicon nanocrystals

The preceding sections have concentrated on SETs fabricated in continuous nc-Si films. In such a film, the GBs are narrow (~1 nm) and if the material is heavily doped, the SETs have comparatively moderate resistance (~100 kΩ or greater) outside the Coulomb blockade region. However, single-electron charging effects tend to be overcome thermally by increasing electron delocalization across the thin GBs. Single-electron charging can persist to higher temperatures in discontinuous, granular silicon films, where higher potential barriers exist between the grains and electrons are localized more strongly. In one of the earliest observations of single-electron effects at room temperature, by Yano et al. (Yano et al., 1995), SETs were formed by nanowires fabricated in ultra-thin (~3 nm), strongly granular, nanocrystalline silicon layers (Fig. 3.24[a]). Here, the grains were only ~1 nm in size. While strong single-electron effects were observed in the *I-V* characteristics, even at room temperature, the tunnel gap resistances were large, leading to a low device current ~10 fA. In a similar room temperature SET design, Choi et al. (Choi et al., 1998) used a thin, discontinuous, PECVD deposited film with 8–10 nm diameter silicon grains, with metal source and drain electrodes separated by a gap of <30 nm.

It is possible to prepare silicon nanocrystals with controlled size and shape, providing a means to form SET islands more precisely. A very promising technique is to grow the silicon nanocrystals using plasma decomposition of SiH_4. This technique has been used to prepare ~8 nm ± 1 nm diameter spherical crystals (Oda et al., 1995; Otobe et al., 1995). The surface oxide on the nanocrystals, ~1.5 nm thick, can form a tunnel barrier isolating the nanocrystals. A number of different configurations of SET designs are possible using these techniques (Dutta et al., 1997, 2000a, 2000b; Nishiguchi and Oda, 2000). Planar devices have been fabricated, where the nanocrystals are deposited between source and drain contacts defined in SOI material, separated by a narrow 30 nm gap (Fig. 3.24[b]). Here, a top-gate supported on a deposited oxide layer is used to control the current, and it is possible to observe single-electron charging effects in the device transconductance up to room temperature (Dutta et al. 2000b). An alternative approach (Fig. 3.24[c]) is to deposit the nanocrystals in a nanoscale hole etched in a silicon dioxide layer, and then top-fill the hole with polycrystalline silicon (Nishiguchi and Oda, 2000).

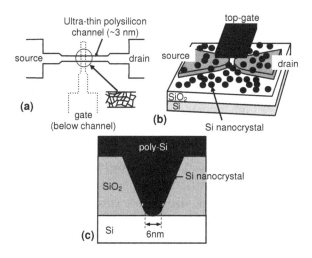

Fig. 3.24 SETs using discrete Si nanocrystals. (a) SET using an ultra-thin, granular nanocrystalline silicon film (Yano *et al.*, 1995). (b) SET with Si nanocrystals deposited in a nanoscale gap between source and drain contacts (Dutta *et al.*, 2000b). (c) Vertical transport device, with Si nanocrystals deposited in a hole etched in a SiO_2 film (Nishiguchi and Oda, 2000).

3.3.6 Comparison with crystalline silicon SETs

We now briefly compare SETs in nc-Si to those in crystalline Si. While nc-Si nanowire and point-contact SETs appear to be superficially similar to crystalline silicon nanowire SETs, the mechanism for formation of the tunnel barriers is different. The tunnel barriers in nc-Si SETs are typically defined by the GBs. While additional disorder effects associated with the device surface, non-uniformity of the dopant distribution or PADOX effects (Section 3.2.1.2) can also occur, these are likely to be less significant. We have seen that careful preparation of the nc-Si film can be used to control the tunnel barrier properties more accurately. This implies that the active regions of the device, i.e. the grains and GBs within the nanowire or point-contact, can be defined precisely by material-processing techniques rather than high-resolution lithography or disorder, and very large numbers of islands can be formed simultaneously over the entire chip area, if necessary. Discrete silicon nanocrystals also provide a means to achieve this. Such a process is an attractive alternative to nanoscale high-resolution lithography over a large area. The SET islands can also be deposited or grown at a convenient stage, greatly increasing the flexibility of the fabrication process.

3.3.7 Electron coupling effects in nanocrystalline silicon

It is possible to investigate electronic interactions between two or more silicon nanocrystals using nc-Si point-contact SETs. Here, the number of nanocrystals taking part in the single-electron transport process can be controlled by varying the dimensions of the point-contact. Furthermore, the interaction of electrons on neighboring nanocrystals can be controlled by tailoring the GB selective oxidation process (Section 3.3.4). It is then possible to operate the device at low temperature as a double- or multiple-quantum dot device, with electron interactions between quantum dots formed by the nanocrystals.

Electrostatic coupling effects have been investigated in great detail at milli-Kelvin temperatures in double quantum dots formed in GaAs/AlGaAs 2-DEG materials (Chapter 2). In these experiments, two gates are used to change the potentials of two quantum dots quasi-independently. A plot of the Coulomb oscillations vs. the two gate voltages shows hexagonal regions of constant electron number on the quantum dots, associated with single-electron interactions between the dots. This forms a 'charge stability' diagram where the total electron number changes by one between neighbouring hexagons. If the quantum dots are strongly tunnel-coupled, then the electron wavefunctions on the two dots can also interact with each other, forming 'quasi-molecular' states analogous to a covalent bond. Resonant tunnelling through these states leads to additional peaks in the device conductance. These states have been observed near ~50 mK temperature in measurements on GaAs/AlGaAs double quantum dots (Blick *et al.*, 1998).

We now discuss electron coupling effects in nc-Si point-contact SETs (Khalafalla *et al.*, 2003, 2004). Our devices consisted of point-contacts

~30 nm × 30 nm × 40 nm in size, fabricated in a ~40 nm-thick heavily doped LPCVD film with grain size from ~10 to 30 nm and ~1 nm-thick amorphous silicon GBs. Two side-gates were used to control the point-contact current. Only a few grains existed within the channel at most, and different grains contributed in varying degrees to the device conduction.

A scanning electron micrograph of the device is shown in Fig. 3.25. By modifying the inter-grain coupling using selective oxidation of the GBs, the electronic interaction between the grains could be controlled. At 4.2 K, only electrostatic interaction, or combined electrostatic and electron wavefunction interaction, could be observed between two quantum dots. Different grains influenced the Coulomb oscillations in different ways, e.g. a single grain or two grains could dominate the Coulomb oscillations, or nearby grains could charge up from one of the contact only, electrostatically switching the oscillations without directly taking part in conduction across the device.

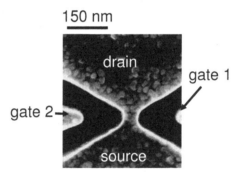

Fig. 3.25 Scanning electron micrograph of a nanocrystalline Si point-contact SET with a 30 nm × 30 nm × 40 nm channel. Multiple nanoscale grains exist in the channel. Electron coupling effects can occur between the grains at 4.2 K.

The GB properties in these devices were controlled by varying the duration of the selective oxidation process, and by subsequent argon annealing. If the device was oxidized at 650–750°C, followed by annealing in argon at 1,000°C, then this created a high and wide GB tunnel barrier (>100 meV), where electrostatic coupling effects dominated. If the device was oxidized only, without annealing, then the GB tunnel barriers remained low (~40 meV) and narrow and the grains were more strongly coupled. These devices showed both electrostatic and electron wavefunction coupling effects.

3.3.7.1 Electrostatic coupling effects

Figure 3.26(a) shows a three-dimensional grey-scale plot of the drain-source current I_{ds} in a device with electrostatically coupled grains at 4.2 K, as a function of the voltages on gate 1 and gate 2 (V_{g1} and V_{g2}, respectively), at V_{ds} = 2 mV. The maximum value of the current (white regions in the plot) is relatively low (I_{ds} = 1.2 pA).

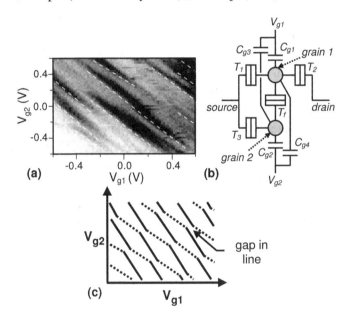

Fig. 3.26 Electrostatic coupling effects in a nanocrystalline Si point-contact SET. (a) Grey-scale plot of the current at 4.2 K, as a function of two gate voltages. V_{ds} = 2mV and the maximum value of I_{ds} = 1 nA (white regions). The Coulomb oscillations form diagonal lines (marked by white dashes) (b) Double quantum dot model. (c) Schematic diagram of lines traced by the Coulomb oscillations in the SET.

A series of lines (marked using white dashed lines) are formed by shifts in the Coulomb oscillation positions as a function of both V_{g1} and V_{g2}. These oscillation lines occur when the single-electron energy levels in the point-contact align with the Fermi energy in the source. As both the gates couple to the grain, the energy of a single-electron level relative to the source Fermi energy depends on a linear combination of the two gate voltages. This leads to the oscillation peaks and valleys tracing diagonal lines across the plot. Switching of the position of the oscillation lines is also observed, which implies an abrupt change in the energy of the corresponding single-electron level. The behaviour can be attributed to single-electron charging of a nearby grain, coupled electrostatically to the dominant grain (Khalafalla *et al.*, 2003).

The characteristics of Fig. 3.26(a) can be understood using the circuit of Fig. 3.26(b). The circuit uses a grain (grain 1), connected to the source and drain by tunnel junctions T_1 and T_2, and coupled capacitively to the two gates by the capacitors C_{g1} and C_{g4}. A nearby grain (grain 2), coupled to grain 1 by the tunnel junction T_f, can be charged with electrons from the source via tunnel junction T_3. Grain 2 is also coupled capacitively to the gates. In such an arrangement, the Coulomb oscillations associated with grain 1 form a series of lines as a function of the two gate voltages, due to the capacitive coupling of grain 1 to *both* gates. The solid lines in Fig. 3.26(c) show this schematically. The electron number on grain 1 differs by one between regions on either side of a line, and this number increases as the gate voltages become more positive. The lines switch in position when the gate voltages overcome the Coulomb blockade of grain 1 and the Coulomb blockade of grain 2 (along the dotted lines) simultaneously, i.e. at the intersection of the solid and dotted lines. At the intersection points, an electron transfers from the source onto grain 2, and this change in charge switches (via the capacitive coupling across T_f) the current through grain 1. Note that there is no direct conduction path from source to drain across grain 2. The overlap between the single-electron oscillation lines in the experimental characteristics is a function of the cross capacitances C_{g3} and C_{g4} between the grains and the gates.

3.3.7.2 Electron wavefunction coupling effects

In contrast to the device characteristics illustrated in Fig. 3.26, which show only electrostatic coupling effects between grains, additional wavefunction coupling effects can be observed in devices that are only oxidized and not annealed. In these devices, the GB tunnel barriers remain low and narrow. In a region (Fig. 3.27[a]) where the Coulomb oscillation lines from two quantum dots QD1 and QD2 (solid and dotted lines, respectively) intersect, the corresponding energy levels from each quantum dot are resonant at two points '1' and '2'. With strong coupling between these levels, additional 'quasi-molecular' states can be formed due to weak GB tunnel barriers. This is shown schematically in Fig. 3.27(b).

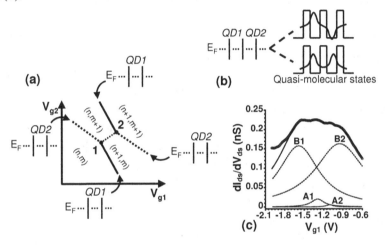

Fig. 3.27 Electron wavefunction coupling effects in a nanocrystalline Si point-contact SET. (a) Position of resonance points ('1' and '2') between energy levels on two quantum dots, shown on a schematic diagram of the Coulomb oscillations vs. gate voltages V_{g1} and V_{g2}. (b) Quasi-molecular states at points '1' and '2'. (c) Device conductance across a line connecting points '1' and '2'. Four peaks (data: thick line) are observed, fitted using four Lorentzian peaks A1, A2, B1 and B2 (fit: dashed lines). These may be attributed to tunnelling through quasi-molecular states (Khalafalla et al., 2004).

Khalafalla *et al.* (Khalafalla *et al.*, 2004) have observed the formation of quasi-molecular states at 4.2 K in an nc-Si point-contact SET, oxidized at 750°C for 30 minutes only. In a measurement of the device conductance across a region where the oscillation lines from two quantum dots intersect, i.e. at the two points '1' and '2' (Fig 3.27[a]), a set of four peaks was observed (Fig. 3.27[c]). These peaks could be fitted using the sum of four Lorentzian peaks. The position of the peaks near the points '1' and '2', where two energy levels in adjacent grains are resonant, and their strongly coupled nature, suggested that they were quasi-molecular states formed by the delocalization of the electron wavefunctions over adjacent tunnel-coupled grains. By comparison, in devices with oxidation and annealing, electron delocalization was inhibited because of the higher and wider GB tunnel barriers.

3.4 Single-Electron Effects in Grown Si Nanowires and Nanochains

Material synthesis techniques have been used to grow crystalline silicon nanowires, and silicon nanochains consisting of a 1-D array of silicon nanocrystals, without the need for high-resolution lithography (Peng et al., 2001; Wu et al., 2004). These nanostructures can, in principle, form building blocks to directly assemble the nanoscale parts of electronic devices (Cui et al., 2003). Silicon nanowires ~10 nm or less in diameter and ~1 μm in length, and silicon nanochains consisting of a series of ~10 nm diameter silicon nanocrystals separated by narrow SiO_2 regions, may be prepared by chemical vapour deposition (Wagner and Ellis, 1964; Cui et al., 2001; Hofmann et al., 2003), or by thermal evaporation of solid sources (Peng et al., 2001).

While the majority of interest in these nanostructures has concentrated on the application of single-crystal silicon nanowires to nanoscale field-effect transistors (Cui et al., 2003; Zhong et al., 2003), Si nanowires and nanochains are also of great interest in the fabrication of single-electron devices (Zhong et al., 2005; Rafiq et al., 2008). In particular, the morphology of silicon nanochains suggests that they may form nanoscale MTJs with the possibility of room temperature single-electron charging. Here, the silicon nanocrystals form the charging islands and the SiO_2 regions form the tunnel barriers.

Zhong et al. (Zhong et al., 2005) have used single-crystal silicon nanowires formed by a chemical vapour deposition reaction to fabricate SETs operating at low temperature. In these devices, the diameter of the crystalline silicon core of the nanowires was only 3–6 nm. The nanowires were synthesized by a vapour-liquid-solid growth process using SiH$_4$, with B$_2$H$_6$ for intrinsic doping. Five nanometre diameter gold nanocrystals acted as catalysts in the growth process. The nanowires were deposited on SiO$_2$-on-Si substrates, with Ni source and drain contacts defined on the nanowires by e-beam lithography. The silicon substrate underneath the nanowire was used as a back-gate. In measurements at 4.2 K, the ~100–400 nm nanowire section between the contacts behaved as a single quantum dot, isolated by tunnel barriers formed at the contacts. The total quantum dot capacitance C ~10 aF, and the energy level spacing in the quantum dot was ~3 meV or less. Single-electron effects in the device persisted up to ~30 K.

Silicon nanochains provide a means to reduce the island size to ~10 nm or less, reducing the island capacitance < 1 aF and raising the maximum temperature for single-electron effects to room temperature. In measurements of large bundles of silicon nanowires and nanochains (Kohno *et al.*, 2005), a Coulomb staircase *I-V* characteristic was observed at room temperature, attributed to single-electron effects in silicon nanocrystals within the bundle. However, the large number of nanowires and nanochains in the bundle complicated a detailed analysis of the results. A device using a single silicon nanochain allows clearer observation of room temperature single-electron effects. In the following, we discuss the fabrication and operation of such a device in detail.

Rafiq *et al.* (Rafiq *et al.*, 2008) fabricated single silicon nanochain devices where multiple step Coulomb staircase *I-V* characteristics were observed at room temperature. Each nanochain 'naturally' defined an MTJ, where the single-electron charging energy $E_C = e^2/2C_{eff}$ for a nanocrystal within the MTJ was ~0.3 eV ~$11k_BT$ at 300 K. The silicon nanochains were synthesized by thermal evaporation of a SiO powder solid source at 1,400°C in a quartz tube furnace (Colli *et al.*, 2007). Argon gas carried the vapour through the tube. Un-doped silicon nanowires and nanochains were synthesized in a cooler part of the furnace at 900–950°C. Depending on the growth conditions, 50–90% of the material formed nanochains, with silicon nanocrystals separated by SiO$_2$ 'necks'.

Figure 3.28(a) shows a TEM image of the nanochains. Nanochain material from the furnace was then dissolved in IPA (0.1 mg/3 ml IPA) using ultrasonic tip agitation, and spun onto a SiO$_2$-on-Si substrate at

5,000 rpm. The silicon nanocrystal diameter in different nanochains varied from <10 nm to ~30 nm, and the separation varied from ~15 nm to 40 nm. The width of the 'necks' varied from approximately the diameter of the silicon nanocrystals to well below this value.

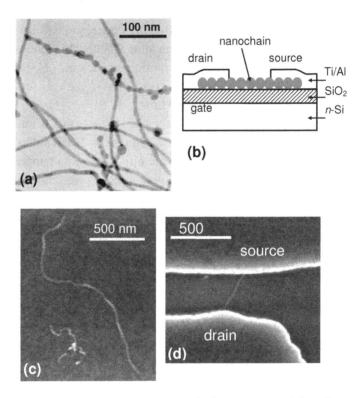

Fig. 3.28 Si nanochain single-electron device. (a) Transmission electron micrograph of Si nanochain material (Reproduced courtesy of A. Colli). (b) Schematic diagram of the device. (c) Scanning electron micrograph of Si nanochains, deposited on a SiO_2 substrate. (d) Scanning electron micrograph of the device.

The silicon nanochain devices were fabricated by defining Ti/Al contacts to selected single nanochains, using e-beam lithography in PMMA resist. Figure 3.28(b) shows a schematic of the device. The devices were defined on SOI material with a ~50 nm-thick SiO_2 capping layer, grown thermally on the top silicon layer of the SOI material. The top silicon layer, 300 nm-thick and doped *n*-type to a concentration of

$1 \times 10^{19}/\text{cm}^3$, formed a conducting back plane for the device, with the potential to form a back-gate.

Initially, an array of Cr/Au alignment marks was fabricated by e-beam lithography on the SiO_2 capping layer. Nanochain material from the furnace, dissolved in IPA (0.1 mg/3 ml IPA) using ultrasonic tip agitation, was then spun onto the sample at 5,000 rpm. Hexamethyl-disilizane vapour treatment of the surface was used to improve surface adhesion. Figure 3.28(c) shows a scanning electron micrograph of a ~20 nm-wide nanochain on a SiO_2 substrate, where the mean separation between nanocrystal centres was ~28 nm. Individual nanochains were then selected with reference to the alignment marks, by SEM inspection. Finally, 20 nm Ti/75 nm Al contacts were defined onto the nanochain using e-beam lithography, after wet etching of the SiO_2 layer around the nanochain in the contact regions. Figure 3.28(d) shows a scanning electron micrograph of a device with a source-drain separation of ~300 nm.

Figure 3.29(a) shows the drain source I_{ds}-V_{ds} characteristics of a nanochain device at 300 K. The source-drain separation in this case was ~180 nm, the nanochain width was ~20 nm and there were seven nanocrystals along the nanochain. The current I_{ds} shows a multiple-step Coulomb staircase (Amman *et al.*, 1989; Grabert and Devoret, 1992). I_{ds} is in the pico-ampere range because of the small size and un-doped nature of the nanochain.

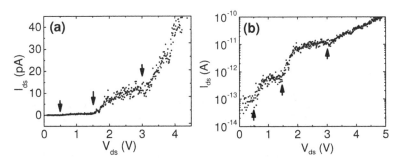

Fig. 3.29 Room temperature *I-V* characteristics of a Si nanochain single-electron device. (a) Coulomb staircase I_{ds}-V_{ds} characteristics on a linear scale. (b) Coulomb staircase I_{ds}-V_{ds} characteristics on a log-linear scale.

Figure 3.29(b) shows I_{ds} plotted on a log scale. Three current steps (arrowed) can be identified in the characteristics, at approximately 0.45 V, 1.5 V and 3 V. A faint, fourth step may exist at ~4 V, (clearer in Fig. 3.29[a]). The threshold voltage for current flow, $V_t \approx 0.45$ V, corresponds to the edge of the Coulomb blockade region.

The Coulomb staircase characteristics may be investigated by single-electron Monte Carlo simulations (the single-electron circuit simulator 'SIMON' is used, see Wasshuber *et al.*, 1997). An N junction MTJ circuit, with equal junction capacitances C and island stray capacitances C_0 for simplicity, can be used to model the nanochains (Fig. 3.30[a]). C_0 corresponds mainly to the nanocrystal to back-plane capacitance. The clear Coulomb staircase suggests a strong asymmetry in the junctions along the MTJ, and this may be modelled by means of a random variation in the tunnel junction resistances R_n, associated with the observed variation in nanocrystal separation.

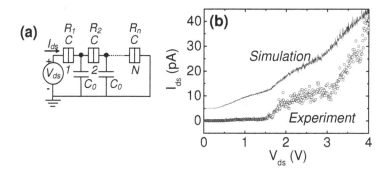

Fig. 3.30 (a) MTJ circuit for the Si nanochain device. Here, C is the tunnel capacitance, C_0 is the island stray capacitance and R_n is the tunnel junction resistance for tunnel junction N. (b) Monte Carlo simulation of the I_{ds}-V_{ds} characteristics at 300 K, for the device of Fig. 3.29. (solid line: simulation, circles: experimental data. The curves are offset by 5 pA from each other for clarity.

Figure 3.30(b) shows the simulation results for an MTJ with $N = 8$ junction and seven nanocrystals. The values of C and C_0 were 0.12 aF and 0.12 aF, respectively, allowing a close match with the experimental values of $V_t \approx 0.35$ V, and the step positions, $V_{ds} = 1.9$ V and 3.6 V. The average tunnel resistance $R_{av} = 6$ GΩ and a 60% random variation in R_n

was applied to individual junctions to obtain a Coulomb staircase. The simulation reproduces V_t and the step positions of the experimental characteristics. It does not, however, reproduce the non-linear increase in current along the staircase, as effects such as a reduction in the tunnel resistances with V_{ds}, or a Schottky-like potential barrier in series with the MTJ, are not considered (Volmar et al., 1998).

A comparison of C and C_0 to the nanocrystal size using the self capacitance of a sphere suggests that the conducting core of the nanocrystal is only ~3 nm in diameter. Furthermore, a significant stray capacitance C_0 is necessary to reproduce both the low observed values of V_t and the wide step widths in the Coulomb staircase. Simulations where C_0 was increased from 0 to 1.2C led to a reduction in V_t, from $V_t \sim (N-1)e/2C = 4.6$ V to $V_t \sim 0.25$ V. This is because, as C_0 is increased, a greater proportion of V_{ds} drops across the first tunnel junction, due to the voltage divider formed by the first junction capacitance C, and the first stray capacitance C_0 in parallel with the equivalent capacitance of the rest of the MTJ (Amman et al., 1989).

The single-electron charging energy of the nanocrystals may be estimated using MTJ theory. For a nanocrystal in the centre of the MTJ, approximating the two halves of the MTJ as semi-infinite capacitive networks, the nanocrystal effective capacitance $C_{eff} = (C_0^2 + 4CC_0)^{1/2}$ (Grabert and Devoret, 1992). For $C = C_0 = 0.12$ aF, $C_{eff} = 0.27$ aF and the single-electron charging $E_c = e^2/2C_{eff} \sim 0.30$ eV $\sim 11 k_B T$ at 300 K. We note that C_{eff} is lower than the total capacitance attached to an island, $C_t = 2C + C_0 = 0.36$ aF. E_c is then higher for a nanocrystal embedded within an MTJ, as compared to a single nanocrystal, leading to an improvement in the Coulomb staircase.

Chapter 4

Single-Electron Memory

4.1 Introduction

Single- and few-electron memories, where information 'bits' are defined by one, or at most a few, electrons, form arguably the most promising application for single-electron effects. As single-electron devices allow control over the charge in a device at the one electron level, this raises the possibility of the definition of digital bits by the presence or absence of a single electron (Likharev, 1988, 1999). If a charging electron can be retained for a useful length of time then the single-electron device can be regarded as a 'single-electron' memory. In a realistic single-electron memory, charge is stored on a dedicated 'memory node', and single-electron effects are used mainly to control the charge. A single-electron memory also requires some means to detect the stored charge, usually by means of a sense amplifier. This part of the memory may be realised by different means, e.g. by using an SET as an electrometer (Nakazato *et al.*, 1993, 1994; Stone and Ahmed, 1998a, 1998b, 2000), a MOSFET as a sense amplifier (Guo *et al.*, 1997; Nakajima *et al.*, 1997; Durrani *et al.*, 1999) or by using a nanocrystalline silicon SET with a memory node embedded within the SET (Yano *et al.*, 1993, 1994).

While the storage charge in a single-electron memory can be as small as one electron, in a practical device it is often more useful to use small packets of electrons per bit, e.g. ten or a few tens (Yano *et al.*, 1999). However, even in such a 'few-electron' memory, it remains possible to control the stored charge with a precision of one electron. As all these devices use the Coulomb blockade effect for the control of the memory charge, an alternative term to describe them is 'Coulomb blockade' memory.

In early experiments on the charging of nanoscale metal islands, Lambe and Jaklevic (Lambe and Jaklevic, 1969) demonstrated that charge could be stored on an electrode one electron at a time. In this work, a thin metal film, consisting of ~10 nm diameter metal islands, was used to form an array of nanoscale capacitors. Thin granular films of Pb, Sn, Bn or In were evaporated on a 7.5–50 nm-thick oxide, deposited on an Al back contact. The metal islands were then oxidized by exposure to air to give a thin tunnelling barrier, and an Al contact was deposited on top. In this device, each island was tunnel-coupled to the top contact and capacitively coupled to the back contact, with an average capacitance ~1 aF. From capacitive measurements of the charging of the islands at 4.2 K, it was demonstrated that the charge build-up was discrete and in the form of single electrons.

A device similar to the structure of Lambe and Jaklevic, where single electrons can be added one at a time, across a small tunnel junction, onto an island connected to a capacitor, is referred to as a 'single-electron box' (Devoret and Grabert, 1992). This device has been discussed theoretically in detail in Chapter 2. The single-electron box may be regarded as the simplest type of single-electron memory. Figure 4.1 shows a circuit diagram of the single-electron box. Here, a tunnel junction with capacitance and resistance C_1 and R_1, respectively, is connected in series with a capacitor C_0. A voltage V is used to add n electrons to the island one-by-one. The island charge is ne.

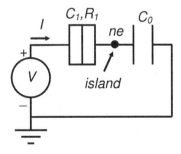

Fig. 4.1 The single-electron box.

Following the analysis of Chapter 2, as V is increased, the average charge Q on the tunnel junction increases until it exceeds a critical charge $Q_c = eC_1/2(C_1 + C_0) = e/2(1 + C_0/C_1)$ and an electron tunnels onto the island. Note that the 'critical charge' is less than e. In a more complex circuit with a number of capacitances connected to the island, we can reduce all these capacitances into an equivalent capacitance C_0. The critical charge equation is then quite general, and may be applied to far more complex circuits. For sufficiently low temperatures such that thermal fluctuations k_BT do not overcome the charging energy of the circuit, the electron number on the island is stable within the range $e(n - 1/2) < C_0V < e(n - 1/2)$, where n is the number of electrons on the island. As the single-electron box allows electrons to be added one-by-one to the island, it can be regarded as a simple memory device. The single-electron box has been realized experimentally by Lafarge *et al.* (Lafarge *et al.*, 1991). Furthermore, part of the circuit developed by Fulton *et al.* (Fulton *et al.*, 1991) for observations of the tunnelling of individual electrons also forms a single-electron box.

4.1.1 Multiple-tunnel junction memory

The single-electron box can be modified by replacing the single tunnel junction by a double or a multiple-tunnel junction (MTJ). Such an 'MTJ memory' often uses an SET as an electrometer to sense the memory-node charge. An MTJ memory in GaAs was used as the basis of the first intentionally-designed single-electron memory cell, operated at 4.2 K by Nakazato *et al.* (Fig. 4.2) (Nakazato *et al.*, 1993, 1994). MTJ traps have also been fabricated using several Al/AlO$_x$/Al tunnel junctions (Dresselhaus *et al.*, 1994), and it is possible to observe the charging of single electrons in these circuits at milli-Kelvin temperatures (Dresselhaus *et al.*, 1994; Krupenin *et al.*, 1997; Matsuoka *et al.*, 1997; Lotkhov *et al.*, 1999). The well-defined parameters in Al/AlO$_x$/Al tunnel junction circuits helps to obtain quantitative agreement between theoretical analysis and experimental data (Matsuoka *et al.*, 1997). A room temperature MTJ single-electron memory has been fabricated using an atomic force microscope (AFM) by Matsumoto *et al.* (Matsumoto *et* al., 2000). Here, the various parts of the circuit were defined using the AFM tip to oxidize sections of a thin Ti film.

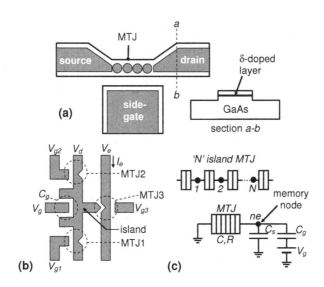

Fig. 4.2 Nakazato single-electron memory in GaAs (Nakazato *et al*, 1993). (a) δ-doped GaAs SET. (b) Layout of memory cell. (c) Circuit diagram.

As the Nakazato memory cell has been used as the basis for many other single-electron memory cells, we shall discuss the design and operation of this device in detail. The cell was fabricated in δ-doped GaAs material grown by metal organic chemical vapour deposition (MOCVD). In this material, conduction occurred mainly in the δ-doped layer, formed by a thin plane of silicon doping at a concentration of $5 \times 10^{12}/cm^2$, lying 30 nm below the surface of the GaAs wafer. The MTJ was defined by etching through the δ-doped layer to form a planar wire with a constricted region 500 nm × 200 nm in area (Fig. 4.2[a]).

Disorder in the doping of the δ-doped layer led to variation in the electron concentration at low temperatures. This formed an MTJ in the constriction at low temperature, where a series of islands was defined by the regions which remained conducting, and tunnel barriers were defined by regions which become insulating (Nakazato *et al.*, 1992). The addition of an in-plane side-gate converted the MTJ into a full SET, and it was possible to observe single-electron oscillations in the MTJ current as a function of the side-gate voltage. The operation of this device is very similar to the silicon nanowire SET, discussed in detail in Chapter 3.

A schematic diagram of the memory cell is shown in Fig. 4.2(b). The total area of this early memory cell was quite large, ~170 μm². The cell uses two MTJ SETs, SET1 and SET2. Here, SET1 is used to control the charging of the memory node, capacitively coupled via a capacitor C_g to a control gate voltage V_g. SET2 is used only as an electrometer, where the charge on the memory node is sensed in the current of the MTJ. To understand the operation of the memory cell, we use the simplified circuit diagram of Fig. 4.2(c), where we do not consider the electrometer formed by SET2. The SET2 gate capacitance and any other stray capacitances are lumped together as the stray capacitance C_s. The MTJ capacitance and resistance are C and R, respectively. For this circuit, the critical charge Q_c is given by (Nakazato et al., 1994):

$$Q_c = \frac{eC}{C_\Sigma}\left(\frac{1+\Delta}{2}\right) \tag{4.1}$$

where $C_\Sigma = C_g + C_s + C$ is the total capacitance connected to the memory node. The parameter $\Delta = (1 - 1/N)(C_\Sigma/C - 1)$, where N is the number of islands in the MTJ (Fig. 4.2[c]). The MTJ remains in Coulomb blockade if the modulus of the charge on the memory node $|Q_m| < Q_c$. Note that if the MTJ is reduced to a single tunnel junction, then $N = 1$, $\Delta = 0$ and $Q_c = eC/2C_\Sigma = e/2(1+ (C_g + C_s)/C)$. This is the same expression as derived for the single-electron box, with $C_0 = C_g + C_s$. In contrast, for large values of N, Eq. 4.1 reduces to $Q_c \approx e/2$. This is a situation equivalent to a single-electron box, where the tunnel junction capacitance $C \ll C_0$.

The memory-node voltage V_m depends upon the gate voltage V_g and the charge ne stored on the memory node, where n is the number of excess electrons on the memory node. V_m is given by:

$$V_m = \frac{e}{C_\Sigma}\left(\frac{C_g V_g}{e} - n\right) \tag{4.2}$$

Figure 4.3(a) shows a plot of this equation, as V_g is swept cyclically. If we assume that initially $n = 0$, then as V_g increases, V_m also increases, along a line with a slope of C_g/C_Σ. There is a corresponding increase in the memory-node charge Q_m, until $Q_m > Q_c$ and V_m becomes greater than the edge of the MTJ Coulomb gap, $V_c = Q_c/C$. The Coulomb blockade on the MTJ is then overcome and an electron transfers on to the memory node from the ground terminal of the MTJ. Now, the electron number n = 1, and the line traced by Eq. 4.2 shifts to the right along the x-axis, dropping $V_m < V_c$. Further increase in V_m leads to additional electrons added to the memory node. Now, if V_g is reversed in direction, initially V_m reduces with n constant and the MTJ remains within the Coulomb blockade. This continues until $V_m = -V_c$, and an electron leaves the memory node, reducing n by one. If V_g is swept cyclically, a hysteresis is traced out. We can then represent the '1' and '0' bits of information by the $+n$ and $-n$ state. In Fig. 4.3(a), these states are $n = +2$ and $n = -2$. In the full memory cell (Fig. 4.2[c]), these states are sensed by the current I_e in the electrometer, SET2.

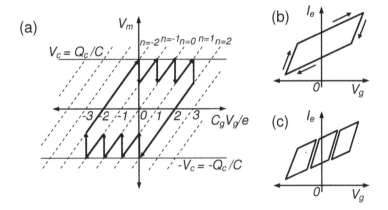

Fig. 4.3 Hysteresis in the Nakazato single-electron memory cell. (a) Memory-node voltage V_m vs. gate voltage V_g. (b) Hysteresis in electrometer current I_e. (c) For a smaller range of V_g, smaller hysteresis loops are seen between the same top and bottom edges.

In the memory cells fabricated by Nakazato *et al.* (Nakazato *et al.*, 1994), $C \sim 3$ aF, $C_g \sim C_g \sim 200$ aF, $C_\Sigma \sim 400$ aF and the electrometer sensitivity was 17nA/V. Figure 4.3(b) shows schematically the hysteresis observed in the electrometer current I_e. The upper and lower legs of the hysteresis were not horizontal, due to the capacitive coupling of the gate voltage to SET2. The hysteresis was 'universal' (Fig. 4.3[c]) i.e. if the gate voltage sweep was reversed at a different voltage, the upper and lower legs of the hysteresis remained the same. This demonstrated that the MTJ controlled the hysteresis. The memory cell could operate up to 4.2 K, with $n = \pm 40$ electrons. However, it was difficult to clearly observe the 'sawtooth' features corresponding to the addition or removal of successive electrons along the upper and lower legs of the hysteresis (Fig. 4.3[a]).

4.2 MTJ Memories in Silicon

An MTJ memory cell operating at 4.2 K has been implemented in SOI material by Stone and Ahmed (Stone and Ahmed, 1998a, 1998b, 2000), using a scaled-down version of the memory cell of Nakazato et al. (Nakazato et al., 1994). Figure 4.4 shows an SEM image and a circuit diagram of a compact, highly-scaled cell, where the cell area was only ~0.5 µm^2 (Stone and Ahmed, 1998b). This is 30 times smaller than initial implementations of the MTJ memory cell in silicon (Stone and Ahmed, 1998a). The cell uses the MTJ formed in a silicon nanowire SET (SET1) (Smith and Ahmed, 1997) to control the charge on the memory node. A second nanowire SET (SET2) is used as an electrometer to sense the charge. A control gate electrode near the memory node is used to form the memory-node capacitor C_m. These various components (SETs, memory node, memory-node capacitor) are all defined in the top silicon layer of the SOI material, forming a planar device. The cell contains all the components of the GaAs cell of Nakazato et al. (Nakazato et al., 1994), defined in silicon in a much more compact fashion.

The device was fabricated in SOI material with a 40 nm-thick top silicon layer, heavily-doped n-type at a concentration of $1 \times 10^{19}/\text{cm}^3$. The fabrication process used a combination of high-resolution e-beam lithography, with a beam diameter of < 10 nm, and trench isolation by RIE in 1:1 SiCl$_4$/CF$_4$ plasma. The SETs consisted of 500 nm-long and 50 nm-wide silicon nanowires, far smaller than the GaAs MTJs used by Nakazato et al.. The small size of the silicon nanowire SETs was crucial to the reduction of the cell area. After the various components were defined by RIE, the device was oxidized in dry O_2 to passivate surface states, reduce the nanowire size further, and improve electrical isolation.

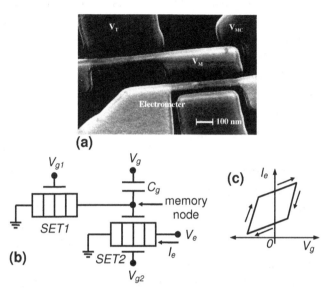

Fig. 4.4 Single-electron memory cell in SOI material. (a) Scanning electron micrograph of cell. (b) Circuit diagram. (c) Hysteresis characteristics. (Reprinted with permission from Stone and Ahmed, *Appl. Phys. Lett.* **73**, 2134 [1998]. Copyright 1998, American Institute of Physics).

The memory cell uses a control gate voltage V_g (Fig. 4.4[b]) to add or remove electrons from the memory node, via SET1. A voltage V_{g1} is applied to the side-gate voltage of SET1 to bias it in a regime where there is a strong Coulomb gap. In contrast, the side-gate voltage V_{g2} of SET2 is used to bias it in a regime where the nanowire current is most sensitive to the memory-node voltage. At 4.2 K, the memory cell operates in a manner similar to the GaAs MTJ cell discussed earlier (Section 4.1.1), and a hysteresis is observed in the current of SET2 as V_g is swept cyclically (Fig. 4.4[c]). Pulsed operation of the cell was also demonstrated, where it was estimated that the memory states were formed by the addition or removal of ~30 electrons.

The MTJ memory of Stone and Ahmed is a fully SET-based circuit, with nanowire SETs used to form both the MTJ and the electrometer. Alternative designs are also possible. Dutta *et al.* (Dutta *et al.*, 1999) have fabricated a memory cell using an MTJ with four islands ~25 nm in diameter, separated by 5 nm-wide constricted regions. The device was defined in SOI material using high-resolution e-beam lithography. The MTJ was connected to a memory node ~45 nm in diameter, and the charge on this node was sensed by a single-island SET. In another few-electron memory design (Takahashi *et al.*, 1998), a scaled MOSFET rather than an MTJ was used to trap ~100 electrons on a memory node. In this design, an SET-based electrometer was retained to sense the small memory-node charge.

In all the above designs, the SET electrometer provides great charge sensitivity and a means to detect each additional electron charging the memory node. However, the lack of gain in an SET implies that the memory charge cannot be sensed with high gain and in a full memory array, so a further sense amplifier may be necessary. However, it is possible to replace the electrometer with a field-effect transistor, e.g. a MOSFET, such that the memory-node charge is sensed with gain in each cell. In such a 'gain-cell' design, each cell has the capability to drive long data lines in an array application. An example of a single-electron memory with gain is the lateral single-electron memory (L-SEM) in SOI material, where each cell uses a SET integrated with a MOSFET (Durrani *et al.*, 1999; Irvine *et al.*, 2000). The charge stored in each L-SEM cell can be as small as ~60 electrons. It is possible to write to each cell using 10 ns pulses, and charge storage persists up to ~77 K. A 3×3 L-SEM cell array has also been developed to demonstrate array operation (Durrani *et al.*, 2000). We discuss the fabrication and operation of the L-SEM in detail in Section 4.5.

4.2.1 The single-electron detector

Stone and Ahmed (Stone and Ahmed, 2000) have demonstrated single electron storage at 4.2 K in a modified version of their MTJ memory cell (Stone and Ahmed, 1998b), using a highly-scaled memory node. This is a good demonstration of the potential of the MTJ memory cell for the fabrication of a true one-electron per bit single-electron memory. The device could also be used to count the number of extra electrons added or removed from the memory node. This silicon-based device may be compared to Al-AlO$_x$-Al MTJ traps, where the charging of single electrons is observed at milli-Kelvin temperatures (Dresselhaus *et al.*, 1994; Krupenin *et al.*, 1997; Matsuoka *et al.*, 1997; Lotkhov *et al.*, 1999).

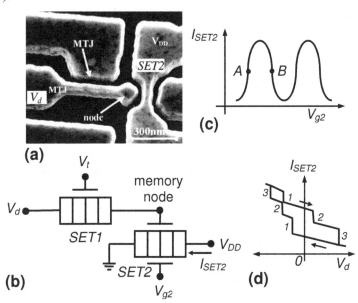

Fig. 4.5 The single-electron detector. (a) Scanning electron micrograph of the device. (b) Circuit diagram. (c) Biasing of SET2. (d) The hysteresis loops in the device correspond to single stored electrons. (Reprinted with permission from Stone and Ahmed, *Appl. Phys. Lett.* **77**, 744 [2000]. Copyright 2000, American Institute of Physics).

Figure 4.5 shows an SEM image and circuit diagram of the cell. Here, the fabricated memory-node diameter was only ~60 nm and the effective conducting size of the memory node was even smaller due to oxidation and surface depletion. A further significant step was the removal of the control gate electrode. The memory-node capacitance was now formed by the capacitance between the memory node and the nanowire of SET2, plus any stray capacitance. The dimensions of SET1 and SET2 were also reduced, with nanowire lengths of only 300 nm and 120 nm, respectively. The nanowire width in both cases was ~35 nm.

The circuit was operated by sweeping the voltage V_d, with constant voltages applied to the SET1 side-gate (V_t), and to the nanowire of SET2

(V_{DD}) and the side-gate of SET2 (V_2). In addition, SET2 was biased using V_{g2} in the single-electron regime and not simply in a region with a strong field-effect. Figure 4.5(c) shows the biasing scheme for SET2. Here, the biasing point could lie either on the positive slope (e.g. biasing point 'A') or on the negative slope (e.g. biasing point 'B') of part of a single-electron oscillation, forming a quasi *n*- or *p*-type device. With the SET biased at point 'A', as V_d was swept cyclically, a hysteresis with sharp steps was observed in I_{SET2}, shown schematically in Fig. 4.5(d). Here, a series of three steps occurs along the upper or lower legs of the hysteresis. Each step corresponded to the addition or removal of single electrons from the memory node.

4.3 Single- and Few-Electron Memories with Floating Gates

It is possible to fabricate single- and few-electron memory cells using 'floating gate' memory nodes to store the charge. Such a memory node may be formed by an array of silicon nanocrystals ~10 nm or less in size (Hanafi *et al.*, 1996; Tiwari *et al.*, 1996a, 1996b; Kim *et al.*, 1999; Hinds *et al.*, 2000; Kapetanakis *et al.*, 2000, 2002; Takahashi *et al.*, 2000a; Normand *et al.*, 2003). Here, each nanocrystal can be small enough such that single-electron charging and quantum confinement effects occur even at room temperature (Kapetanakis *et al.*, 2002; Pace *et al.*, 2005).

Alternatively, a single, ultra-small (~10 nm scale) floating gate may be defined lithographically, such that it is possible to store single electrons in the memory cell (Guo *et al.*, 1997; Nakajima *et al.*, 1997a, 1997b). The charge on the floating-gate memory node may be sensed by a field-effect transistor, forming a configuration very similar to a FLASH memory cell. A background charge insensitive memory design has also been proposed (Likharev and Korotkov, 1995; Chen *et al.*, 1997; Likharev, 1999; Sakamoto *et al.*, 1999; Sunamura *et al.*, 1999), where the single-electron oscillations in an SET are used to detect the charge. Such a memory can be insensitive to 'offset' charge fluctuations in the environment of the SET.

We consider first silicon nanocrystal floating gate memory cells, developed originally by Tiwari *et al.* (Tiwari *et al.*, 1996a). These memory cells are analogous to non-volatile FLASH memory cells, where charge is stored on a discontinuous floating gate formed by a layer of silicon nanocrystals, rather than a single, continuous floating gate. Figure 4.6(a) shows a schematic diagram of such a memory cell. The silicon nanocrystals, usually 1–10 nm in size, can be grown by a variety of means (Hanafi et al., 1996), e.g. by using a silicon-rich oxide, or by high-density silicon ion-implantation into an oxide. Size-controlled nanocrystals can also be created by plasma decomposition of SiH_4 (Hinds et al., 2000). The nanocrystals are then sandwiched in the gate-stack of a silicon MOSFET, separated from the channel by a thin tunnelling oxide layer and from the gate by an additional, thicker control oxide layer.

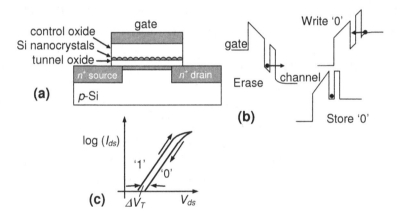

Fig. 4.6 Si nanocrystal floating-gate memory. (a) Schematic diagram of cell. (b) 'Write' and 'erase' operations. (c) Hysteresis characteristics.

Electrons can be injected into the nanocrystals by direct tunnelling from the channel across the tunnel oxide, using a large positive voltage on the gate (Fig. 4.6[b]). This writes a '0' into the cell, which is sensed by a shift ΔV_T in the threshold voltage of the MOSFET (Fig. 4.6[c]). The charge can be erased by applying a large positive gate voltage, allowing electrons to tunnel back into the channel (Fig. 4.6[b]). We note that these memory cells are 'single-electron' only in the sense that one electron per nanocrystal can be stored, and single-electron effects can be inherent in this process. However, the devices have the potential to be scaled-down to a degree such that only a few nanocrystals form the floating gate and only a few electrons are stored per cell.

In conventional FLASH memory cells, the thick oxide layers used to isolate the floating gate imply that a high voltage is necessary to transfer electrons on to the gate. This process occurs through Fowler-Nordheim (FN) tunnelling, or by hot-electron injection. The high electric field and current in the FN tunnelling process generates traps, degrading the oxide over numerous operation of the cell. This leads to an increasing leakage current over many cell operations, limiting the lifetime of the memory.

The problem becomes even more severe in a scaled FLASH memory cell, where the charge stored is small and any leakage path can quickly discharge the cell. By contrast, in a nanocrystal memory cell, the electric field can be smaller and any leakage paths affect only a small number of nanocrystals, as there is a large oxide barrier of ~3 eV between nanocrystals. Charge is retained in the remaining nanocrystals and it is difficult to fully discharge the cell. A typical example of such a memory cell, fabricated by Hanafi *et al.* (Hanafi *et al.*, 1996), uses 3 nm diameter silicon nanocrystals at a density of $1 \times 10^{12}/cm^2$. The nanocrystals are separated from each other by ~7 nm. The lower oxide layer is only ~1 nm thick. If a charge corresponding to one electron per nanocrystal is stored, then the threshold voltage shifts by 0.35 V, which can change the underlying MOSFET threshold current by 4–5 orders of magnitude. The retention time of such a memory cell can be ~10^5 s.

Considerable refinement of these cells is possible, by using silicon nanocrystals implanted at very low energy, 1 keV, to obtain a 2-D array of nanocrystals at a small distance from the channel (Kapetanakis *et al.*, 2000, 2002; Normand *et al.*, 2003). In the work of Normand *et al.* (Normand *et al.*, 2003), the silicon nanocrystal average size was only ~2 nm, implanted at 1 keV at a density of $2 \times 10^{16}/cm^2$ into gate oxides only ~7 nm thick. Annealing in dilute oxygen narrowed the nanocrystal size distribution, improved the quality of the oxide and thickened the control oxide. Charge could be retained on the nanocrystals for ~11 hours. In similar devices, is was also possible to observe a clear staircase along the legs of the current hysteresis, caused by single-electron charging of the nanocrystals at room temperature (Kapetanakis *et al.*,

2002). In other work (Hinds *et al.*, 2000), the nanocrystals were created by plasma decomposition of SiH_4 and the nanocrystal size was controlled precisely. Alternatively, it is also possible to use an SET rather than a MOSFET to detect the nanocrystal charge, reducing the device area considerably (Takahashi *et al.*, 2000a).

We now consider floating-gate memory cells where a single, ultra-small floating gate is defined lithographically, such that it is possible to store single electrons per memory cell (Guo *et al.*, 1997; Welser *et al.*, 1997; Nakajima *et al.*, 1997a, 1997b, 1999a). A schematic of this type of device, fabricated usually with e-beam lithography, is shown in Fig. 4.7(a). More recently, nano-imprint lithography has been used to define these devices, of considerable interest for large-scale fabrication (Wu *et al.*, 2003). The devices use oxidized silicon nanowires, formed in the top silicon layer of SOI material. A square polysilicon or amorphous silicon gate is then defined above the nanowire. The dimensions of this gate can be further reduced by wet etching, or by oxidation, e.g. Guo *et al.* used oxidation to reduce the dimensions to only ~7 nm square. A further layer of oxide, with a control gate on top, completes the device.

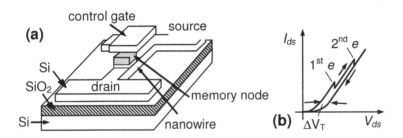

Fig. 4.7 Memory cell with a nanoscale floating gate. (a) Schematic diagram. (b) Hysteresis characteristics.

In a manner similar to the nanocrystal memories discussed above, charge is usually injected from the nanowire channel using a control gate (Fig. 4.6[b]). This is then detected in a threshold voltage shift in the nanowire current as a function of the control gate voltage. The ultra-small size of the island leads to large single-electron charging energies comparable to or greater than the thermal fluctuations $k_BT = 26$ meV at room temperature.

In the device of Guo et al., with a 7 nm square memory node, the charging energy was very large, ~3.6 V. These large charging energies imply that it is possible to control the addition of single electrons onto the floating gate, even at room temperature. Figure 4.7(b) shows the nanowire current I_d vs. the control gate voltage V_g schematically, e.g. similar to the experimental results of Nakajima et al. (Nakajima et al., 1997a). As V_g is increased, small shifts in the threshold voltage occur, leading to small peaks in I_d. Each peak corresponds to the addition of a single electron to the floating gate, and the associated threshold voltage shift is of the same width. If V_g is reduced, a hysteresis is observed, corresponding to the charge stored on the floating gate. The cell can then be operated with a precisely known number of electrons, as small as one electron, at room temperature. It is also possible to operate the nanowire as an SET at lower temperatures, forming an SET-sensed single-electron memory (Nakajima et al., 1997b).

Single or small numbers of electrons on a floating gate can be sensed with maximum sensitivity by an SET, in a manner similar to the MTJ trap with an SET electrometer (Nakazato et al., 1993). However, as the SET may be extremely sensitive to background 'offset' charge fluctuation (Devoret and Grabert, 1992), read-out of such a memory can be unreliable. Likharev and Korotkov (Likharev and Korotkov, 1995; Likharev, 1999) have proposed a background-charge insensitive memory design (Fig. 4.8), where single-electron oscillations in the SET are used to both control the amount of charge added to a floating-gate memory node, and to sense this charge.

The memory requires the detection of single-electron oscillations of a given period and any 'offset' charge only shifts the phase of the oscillations and not the period. This type of memory, originally realized using Al/AlO$_x$ SET sensing at low temperature (50 mK–3 K) (Chen et al., 1997; Sunamura et al., 1999), has also been realized in silicon (Sakamoto et al., 1999). Similar cells in GaAs have been shown to operate up to 77 K (Yoo et al., 1999).

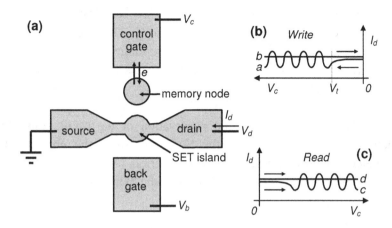

Fig. 4.8 Background charge insensitive memory. (a) Schematic diagram of memory cell. (b) 'Write' operation. (b) 'Read' operation.

A background-charge insensitive memory cell is shown schematically in Fig. 4.8(a). The cell consists of a floating memory node between an SET and a control gate. A back-gate is used to trim the SET characteristics. The memory cell is operated using the control gate voltage V_c and the back-gate voltage V_b. To write electrons to the memory node, V_c is swept negative (Fig. 4.8[b]). A voltage is also applied to the back gate, $V_b = -\alpha V_c$, to compensate the effect of the control gate on the SET characteristics.

Here, α is the ratio of the capacitance between the SET island and the control gate, to that between the SET island and the back-gate. As V_c is swept negative (curve 'a'), initially there is no additional charge on the memory node, and the net gating effect on the SET island remains compensated. The SET current then remains constant. However, after a threshold voltage V_t, electrons are added by FN tunnelling to the memory node. A net gating effect now occurs on the SET, and as V_c is swept further, single-electron oscillations occur in the SET current. This is a signature of the charging of the memory node and forms the 'write' operation. Sweeping back (curve 'b') does not show any oscillations as the threshold voltage magnitude increases. To read the charge on the memory node, a positive voltage is applied to V_c. The detection of single-electron oscillations within a known voltage range implies the presence of a '1' (curve 'c'). However, the 'read' process is destructive, and the observation of the same number of periods as in the writing operation fully discharges the memory node. A repeat of the 'read' operation (curve 'd') does not lead to the observation of the oscillations. Any background 'offset' charge on the SET only shifts the phase of the single-electron oscillations, and not the period. As the observation of single-electron oscillations is used to detect the memory-node charge, the memory is background-charge independent.

4.4 Large-scale Integrated Single-Electron Memory in Nanocrystalline Silicon

A large-scale integrated (LSI) single-electron memory, operating at room temperature and capable of integration at the Gbit–Tbit scale, has been developed by Yano and co-workers (Yano et al., 1993, 1994, 1996a, 1998, 1999; Ishii et al., 1997). Each cell in the memory is formed by an ultra-thin film transistor, fabricated in a nanocrystalline silicon film on average only ~3 nm thick. The nanocrystalline silicon film is strongly granular, with crystalline silicon grains only a few nanometres in size. Due to the very small size of the grains, strong single-electron and quantum confinement effects can occur on the grains, even at room temperature. A schematic of the device is shown in Fig. 4.9(a).

The nanocrystalline silicon film forms the device channel between drain and source lines 'DL' and 'SL', respectively. The channel length and width is 200 nm and 100 nm, respectively (Plan view, Fig. 4.9[b]). A word-line 'WL', supported on a 30 nm-thick gate oxide, forms a gate over the entire channel area. A voltage V_{wl}, applied to 'WL', is used not only to form a percolation conduction path through a series of grains in the nanocrystalline silicon channel, but also to control charge stored on a grain isolated from the percolation path. This grain forms the memory node in the cell. We note that here, the memory node is embedded within the channel area of the sense amplifier formed inherently by the cell.

Both the percolation path and the memory node lie in the channel and are controlled simultaneously by the word-line gate.

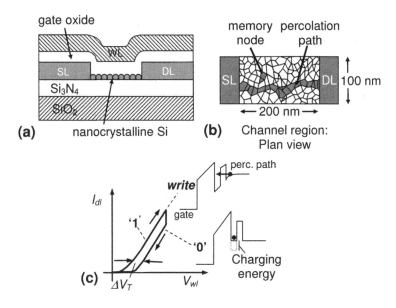

Fig. 4.9 Yano single-electron memory (Yano *et al.*, 1999). (a) Cross-section of the memory cell. (b) Plan view of the memory cell. (c) Hysteresis characteristics.

Figure 4.9(c) schematically shows the behaviour of the drain line current I_{dl} as a function of V_{wl} at a constant drain-source voltage. As V_{wl} is increased, I_{dl} begins to flow when a percolation path forms between the source and drain. This corresponds to a low threshold voltage with zero charge stored within the device. In these devices, the percolation path at its narrowest is estimated to be ~10 nm, forming a bottleneck region. However, the increase in V_{wl} also lowers the potential of other grains, isolated from and near to the bottleneck. As V_{wl} increases further, an electron tunnels from the percolation path onto a nearby grain. This electron forms the stored charge in the memory. The potential of the electron shifts the threshold voltage by ΔV_T, with a corresponding drop in I_{dl}. The memory cell is now in the '0' state. If V_{wl} is reduced, the cell remains in the '1' state. Values of V_{wl} up to 15 V are needed to operate the cell. The capacitance of the memory node was estimated to be ~ 2 aF.

The corresponding single-electron charging energy E_c ~40 meV, greater than $k_BT = 26$ meV at room temperature.

Figure 4.10(a) shows the organization of these memory cells into an array. The array consists of a series of word, drain and source lines. A single source line is shared by two cells, reducing the cell area to $6F^2$, where F is the feature size. The drain and source lines form the data line for the cells. The biasing of the array for the 'read', 'erase' and 'write' operations is shown in Fig. 4.10(b), for four cells. Two target cells, cell 'A' and cell 'B', will be considered. To 'read' the target cells, 2.5 V is applied to the common word-line, and 2.5 V to the respective drain lines.

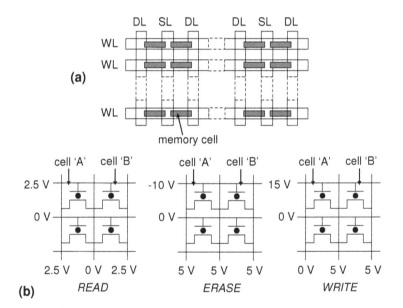

Fig. 4.10 Yano single-electron memory array (Yano et al., 1999). (a) Array plan. (b) 'Read', 'write' and 'erase' operations.

The common source line is biased at 0 V. The memory states of the cells can then be sensed in their respective drain line currents. The array also uses an 'erase' operation before every write, to prevent the possibility of multiple writing of states. This uses a voltage of −15 V, applied to the common word-line, to remove any stored electrons in the cells. To 'write' to a given cell, e.g. cell 'A', but not write to cell 'B',

15 V is applied to the common word-line, 0 V to the drain line of cell 'A', 5 V to the common source line and 5 V to the drain line of cell 'B'. The bias between the gate (word-line) and the channel is then high enough to write an electron only into cell 'A'.

It is possible that a '1' is not written in a single 'write' operation. The array therefore also uses a 'verify' operation (Ishii *et al.*, 1997a), where after every 'write', the cell is 'read' to check if a '1' has been written. If a '1' has not been written, the 'write' operation is repeated. It was determined that the memory cells could be operated with five stored electrons using 'verify', and with ten stored electrons without using 'verify'. The write/erase time was typically ~10 μs and the retention times varied from 1 hour to a month.

Yano *et al.* (Yano *et al.*, 1998, 1999) have used these basic designs to fabricate a 128 Mbit memory circuit using 0.25 μm technology, where each source line had the two adjacent data lines above and below it, and the cells were defined vertically in a 3-D integrated scheme, reducing the cell area to $2F^2$ per bit (Ishii *et al.*, 1997b). Here, the cell size was only 0.145 μm²/bit. Sixty-four cells shared a local data line, and one block, including the 'upper' and 'lower' data lines, consisted of 128 cells. One global data line shared 128 blocks. Read, write and erase were performed simultaneously using a word-line with 8,000 cells. The designed read access time of the circuit was 20 μs. The peripheral circuit used 0.4 μm CMOS technology, except for the larger high-voltage transistors necessary for the write/erase operation. However, due to fabrication difficulties, half of the cells showed no current.

4.5 Few-Electron Memory with Integrated SET/MOSFET

We now discuss in detail the fabrication and operation of the lateral single-electron memory (L-SEM), an integrated SET/MOSFET memory cell in SOI material (Durrani *et al.*, 1999; Katayama *et al.*, 1999; Irvine *et al.*, 2000; Mizuta *et al.*, 2001). The memory cell uses a silicon nanowire SET (Smith and Ahmed, 1997), fabricated in the top silicon layer of the SOI material, to trap small numbers of electrons (as few as ~60) on a memory node. The Coulomb blockade effect in the SET leads to two charge states on the memory node. These states are then sensed, with gain, by a MOSFET using a channel lying in the substrate of the SOI material. The use of the Coulomb blockade effect allows the voltage on the memory node to be controlled at the 0.1 V scale, allowing control over a very small charge, consisting of a few tens of electrons, with memory node storage capacitance ~10^{-16} F.

The L-SEM is not a 'single-electron' memory in the sense that the presence or absence of one electron defines the 'bits'. However, the memory cell is *controlled* by an SET in Coulomb blockade, and electrons are added one-by-one to the SET during memory operation. The charge stored on the memory node can be as small as ~60, a large improvement on the ~50,000 or more electrons stored in complementary metal oxide semiconductor (CMOS) dynamic random access memory (DRAM). It is also possible to scale the design of this type of memory to the one electron per bit level, e.g. by using an SET rather than a MOSFET to sense the charge (Stone and Ahmed, 2000).

The silicon nanowire SET used in the L-SEM is formed by an etched nanowire ~50 nm × 50 nm or less in cross-section. The nanowire inherently forms a MTJ, which provides the single-electron charging effect used as the basis of the cell operation (Chapter 3). Two in-plane side-gates are used to control the nanowire current. In SETs of the type used in this work, it is usually possible to observe gate-controlled Coulomb blockade and single-electron current oscillations up to ~77 K.

We note that, in SETs using narrower silicon nanowires only ~10 nm × 10 nm in cross-section, it is possible to observe single-electron effects even at room temperature (Ishikuro *et al.*, 1996). The L-SEM memory states, controlled by the Coulomb gap in the SET, can be observed in the form of a hysteresis in the MOSFET channel current at 4.2 K. Current steps exist along the hysteresis characteristics, which may be attributed to single-electron oscillations in the SET. The memory-node area, and therefore the stored charge, can be reduced by using a split-gate L-SEM design. This cell uses a dual-gate MOSFET with a 'memory node' gate and an additional 'outer gate'. Memory cells with 1 μm × 1 μm and 1 μm × 70 nm memory nodes have been demonstrated, and operated using pulse widths down to ~10 ns.

A 3 × 3 L-SEM cell array has also been developed using the split-gate L-SEM cells (Durrani *et al.*, 2000). Here, the MOSFET outer gate is used for cell selection. A combination of read/write pulses is necessary to operate the cells in the array. Here, the measured states are separated by ~1,000 electrons for the 1 μm × 1 μm memory node cell and by ~60 electrons for the 1 μm × 70 nm memory node cell. SET-controlled operation persists up to a temperature of 65 K. The operating temperature is somewhat low due to the rather large SET size. However, it may be possible to increase this temperature by using smaller SETs with better operating temperature.

The experimental L-SEM devices discussed here are fabricated using e-beam lithography. However, the fabrication process is CMOS-compatible and the extension of the design to a much larger array is possible with high-throughput sub-50 nm resolution optical lithographic processes.

4.5.1 Silicon nanowire SETs for L-SEM application

For the L-SEM application, the primary SET design was the silicon nanowire SET in wafer-bonded, crystalline SOI material. An alternative version of this device was fabricated in polycrystalline SOI material (Irvine *et al.*, 1998), which may have the advantage of increasing the flexibility of the fabrication process. In this section, we discuss these devices briefly with a view to their operation in the L-SEM. Further details regarding the devices may be found in Chapter 3.

The crystalline, wafer-bonded SOI material, prepared at the Laboratoire d'Electronique de Technologie de l'Information (LETI), Grenoble, France, consisted of a lightly-doped *p*-type silicon substrate with a 40 nm-thick thermally-grown buried oxide layer and a 40 nm-thick top silicon layer. The top silicon layer was heavily doped *n*-type to a doping concentration of $1-2 \times 10^{19}/cm^3$ using arsenic ion implantation.

The polycrystalline SOI material was prepared at IMEC, Leuven, Belgium, in the following manner. A 50 nm-thick amorphous silicon layer was deposited at 550°C on 10 nm-thick gate quality oxide, grown thermally on a standard, lightly-doped (*p*-type, $5 \times 10^{14}/cm^3$) silicon substrate. The amorphous silicon was then heavily doped *n*-type to a doping concentration of $3–4 \times 10^{19}/cm^3$, and annealed to form polycrystalline silicon with a grain size of ~20 nm.

The fabrication process and the basic device geometry of the SET in both types of materials were similar. The nanowire and adjacent side-gates were defined in the top silicon layer. High-resolution e-beam lithography using PMMA resist, followed by etching in $CF_4/SiCl_4$ reactive-ion plasma, was used to fabricate the device. In the crystalline SOI SETs, wire widths from 30 to 70 nm and wire lengths of 1 µm were used. In the polycrystalline SOI material, SETs with wire widths of 50 to 60 nm and wire lengths ~1 µm were used. An oxide layer ~20 nm thick was grown thermally to thin the wire and passivate the silicon surfaces.

Figure 4.11 shows a scanning electron micrograph of a polycrystalline SET with a ~20 nm-thick oxide layer. Here, the unconsumed silicon cross-section in the wire is approximately $20 \text{ nm} \times 30 \text{ nm}$, roughly comparable with the average grain size. This implies that the wire is likely to consist of a near-one-dimensional chain of polycrystalline silicon grains.

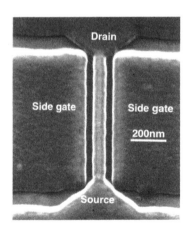

Fig. 4.11 Polycrystalline Si nanowire SET.

Although single-electron charging effects were observed in both types of devices, there were significant differences in the detailed electrical characteristics. In the following, we compare the electrical behaviour of either type of device.

4.5.1.1 Nanowire SET in crystalline silicon

This device uses a nominally uniform nanowire in crystalline Si, sufficiently small in width such that any fluctuations in the doping density create significant variation in the width of the surface depletion regions. Thus the conducting pathway along the nanowire is of variable width. Non-uniformities in the density of surface states can contribute further to this. Applying a negative side-gate voltage V_{trim} increases the depletion width, causing localized necking in the conductive pathway along the nanowire where the local dopant density is lowest, and/or where the nanowire is narrowest. In this case, the wire forms a chain of conducting islands separated by depletion barriers, i.e. an MTJ where single-electron charging can occur.

Figure 4.12 shows the electrical characteristics of a 50 nm-wide (before oxidation) and 1 μm-long nanowire SET at 4.2 K. In Fig. 4.12(a), the nanowire current-voltage (*I-V*) characteristic at 4.2 K is plotted for a series of side-gate voltages. At positive V_{trim}, the wire conduction is almost linear, corresponding to a situation where the nanowire channel is approximately continuous. As V_{trim} is decreased, the non-linearity becomes more pronounced, and a zero-current Coulomb gap is observed. More negative values of V_{trim} widen the Coulomb gap further until the wire is pinched off completely, at $V_{trim} \approx -10$ V.

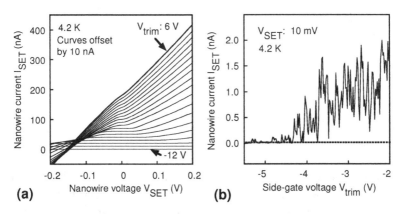

Fig. 4.12 The L-SEM: Electrical characteristics of the SET. (a) Drain-source *I-V* characteristics. (b) Coulomb oscillations in the gate characteristics.

In Fig. 4.12(b) the nanowire current is plotted against the side-gate voltage at a fixed nanowire voltage. Complex, reproducible single-electron oscillations are seen in the current. With narrower wires, it is possible to observe these effects at temperatures well above 77 K. The electrical characteristics of these devices have been modelled qualitatively as a combination of MTJ and field-effect transistor behaviour (Müller et al., 1999, 2000).

4.5.1.2 Nanowire SET in polycrystalline silicon

Figure 4.13 shows the electrical characteristics of a 50 nm-wide (before oxidation) and 1 µm-long polycrystalline silicon nanowire SET at 4.2 K. Figure 4.13(a) shows the nanowire I-V characteristics, plotted as a function of the side-gate voltage V_{trim}. A multiple-step Coulomb staircase is seen, modulated periodically by the side-gate voltage.

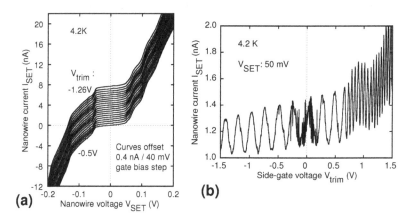

Fig. 4.13 The L-SEM: Electrical characteristics of a polycrystalline Si nanowire SET (a) Drain-source Coulomb staircase I-V characteristics. (b) Coulomb oscillations in the gate characteristics.

In this particular device it is difficult to reduce the Coulomb gap to zero using the side-gate, in a manner similar to the crystalline silicon nanowire SET. This suggests an MTJ where the charging of some of the islands cannot be controlled by the gate voltage. Single-electron oscillations are observed in the drain-source current as a function of side-gate voltage (Fig. 4.13[b]). From −1.5 V to 0.5 V, the oscillation period is 230 mV, decreasing sharply to 50 mV at higher voltage. The effective gate-island capacitances $C_g = e/V_g$ are 0.7 aF and 3.2 aF, respectively.

The differences in electrical behaviour between polycrystalline and crystalline silicon nanowire SETs may be attributed to the strong influence of the polycrystalline silicon GB states in the former device. The GBs contain a high concentration of defect states, segregated dopants and other impurities, which may pin the Fermi level, strongly attenuating gate action for part of the range of applied gate voltage and creating a change in the single-electron oscillation period (Irvine et al., 1998).

Fermi level pinning in some of the islands in the MTJ may also explain the lack of full modulation of the Coulomb gap. The change in the single-electron oscillation period has also been attributed to the formation of an inversion layer underneath the gate region at positive gate voltages (Tan et al., 2001a). This would change the effective gate-island capacitance and hence the oscillation period.

4.5.1.3 Potential for mass fabrication

The crystalline and polycrystalline silicon nanowire SETs are robust, silicon-based MTJ devices which can be fabricated using standard CMOS processing techniques. The oxidation stage in the fabrication of the device not only thins the SET nanowire, it also passivates the defect states and considerably reduces any 'offset' charge switching of the device (see Chapter 2). The use of MTJs further helps in reducing the effect of offset charges. In addition, as CMOS devices are scaled into the sub-50 nm regime, it may become possible to fabricate LSI SET circuits.

In the majority of our devices, the minimum feature size is still somewhat large at ~50 nm, although this is reduced further using oxidation. As the feature size is reduced to ~10 nm, room temperature operation may become possible.

The polycrystalline silicon nanowire SET, fabricated in deposited polycrystalline silicon layers, can be flexibly incorporated in a CMOS fabrication process. However, compared to the crystalline SET, the Coulomb gap may be difficult to vary over a wide range with side-gate voltage, and appears to depend more on the material characteristics. This is an advantage if only a single Coulomb blockade voltage is required. If a variable Coulomb blockade voltage is needed, the use of polycrystalline silicon with lower GB defect state density, or nanocrystalline silicon with grains ~10 nm in size, may improve the characteristics (Chapter 3).

We note that nanocrystalline silicon 'point-contact' SETs operating at room temperature have been demonstrated (Tan *et al.*, 2003). The ultra-thin polysilicon SET of Yano *et al.* (Yano *et al.*, 1995) also operates at room temperature, although the low drain-source conductance of the device implies that the charging current in L-SEMs using such an SET may be comparatively low. These devices are clearly of great potential for room temperature operation of single- and few-electron memories.

4.5.2 Single-gate L-SEM

We first discuss the fabrication and operation of the single-gate L-SEM, fabricated in wafer-bonded crystalline SOI material (Durrani *et al.*, 1999). This device has been theoretically analysed in detail by Katayama *et al.* (Katayama *et al.*, 1999). Here, a nanowire SET is connected to a memory node defined in the upper silicon layer of the SOI material (Fig. 4.14). The memory node forms the gate of an *n*-channel MOSFET, where the MOSFET source, drain and channel regions lie in the substrate silicon of the SOI material. We characterize a memory effect in the cell at 4.2 K, controlled by the SET and sensed in the MOSFET current.

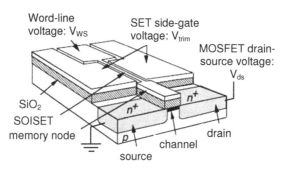

Fig. 4.14 Schematic diagram of the L-SEM.

We explain the operation of the memory cell following the analysis of Katayama et al. (Katayama et al., 1999). A simplified circuit diagram of the cell is shown in Fig. 4.15(a). Here, the MTJ is represented by a double tunnel junction circuit, with capacitances and tunnel resistances C_1, C_2 and R_1, R_2, respectively. This is equivalent to a situation in an MTJ where the free energy of the system is a maximum, and a charge is placed on one of the central islands, as shown in Fig. 4.15(b). The MTJ connects the memory node to a word-line, with an applied voltage V_w. The word-lines form the rows in an L-SEM array (Section 4.5.4). C_s represents the memory node to MOSFET source capacitance, and C_d represents the memory node to MOSFET drain capacitance. In this simplified circuit, a data line is connected to the MOSFET drain only, with an applied voltage V_d. As we will see (Section 4.5.4), L-SEM cells in an array use two data lines connected to both the drain and source of the MOSFET. During a write operation, these data lines are connected in common, with identical voltages. The two data lines then form a single column in the L-SEM array.

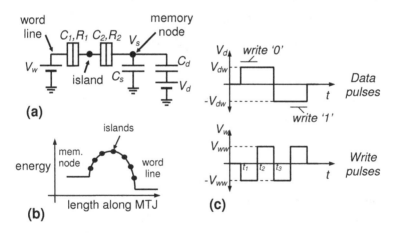

Fig. 4.15 L-SEM operation. (a) Circuit diagram. (b) Energy diagram across the MTJ. (c) Data and write pulses.

The write operation uses the pulses shown in Fig. 4.15(c). To write a '0', a positive data pulse $V_d = V_{dw}$ is applied to the data line. At the same time, a write pulse consisting of a negative voltage pulse $V_w = -V_{ww}$ is applied to the word-line, e.g. between times t_1 and t_2. If the difference voltage $V_w - V_d$ is large enough to overcome the Coulomb blockade voltage V_c in the MTJ, then electrons are written to the memory node from the word-line. The use of two pulses allows selection of a given cell in an array. To write a '1', pulses of opposite polarity are used. The memory-node voltage V_s as a function of V_w and V_d, for a charge Q_s stored on the memory node, is given by:

$$V_s = (C_{MTJ}V_w + C_d V_d + Q_s)/C_\Sigma \qquad (4.3)$$

where $C_\Sigma = C_{MTJ} + C_d + C_s$ and $C_{MTJ} = C_1 C_2/(C_1+C_2)$. If $|V_w - V_s| > V_c$, then the memory node charges until $|V_w - V_s| = V_c$. For a write '0' operation, this condition, combined with Eq. 4.3, gives the required value of V_{ww} for a given value of V_{dw} and memory-node charge Q_{s0}:

$$V_{ww} = \frac{C_\Sigma V_c - C_d V_{dw} + |Q_{s0}|}{C_s + C_d} \qquad (4.4)$$

In the standby condition, when V_w and V_d are zero, Eq. 4.4 gives the charge retained by the cell, $|Q_{s0}| = C_\Sigma V_c$. To avoid destruction of the data in non-selected cells, the following conditions are necessary:

$$(2(C_s + C_d)V_{ww} + C_d V_{dw})/C_\Sigma < 2V_c \qquad (4.5)$$

$$((C_s + C_d)V_{ww} + 2C_d V_{dw})/C_\Sigma < 2V_c \qquad (4.6)$$

The write time, i.e. the time required to charge-up the memory node can be approximated as:

$$t_{write} = R_{MTJ}C_\Sigma \qquad (4.7)$$

Here, R_{MTJ} is the total MTJ resistance. Finally, by using the 'orthodox theory' of single-electron tunnelling (Dresselhaus et al., 1994), Katayama et al. (Katayama et al., 1999) derived a simple expression for the retention time of the cell, as a function of V_c:

$$t_{1/2} \approx \frac{RC_s k_B T}{eV_c} \exp\left(\frac{eV_c}{2k_B T}\right) \qquad (4.8)$$

where $t_{1/2}$ is the time taken to lose half the stored charge Q_{s0}, $R_1 = R_2 = R$ and it is assumed that $C_1 = C_2 \ll C_s$.

4.5.2.1 Single-gate L-SEM fabrication and characterization

The single-gate L-SEM cell (Fig. 4.16) was fabricated in SOI material with a 40 nm-thick upper silicon layer, on top of a 40 nm-thick buried-oxide layer. The substrate was lightly-doped p-type. The upper silicon layer was heavily doped n-type by arsenic implantation (8×10^{13} cm^{-2} at 35 kV). The substrate channel region was implanted p-type with boron (2×10^{12} cm^{-2} at 55 kV and 5×10^{12} cm^{-2} at 75 kV). The substrate source/drain contact regions, separated by 8 µm, were heavily doped n-type by phosphorous implantation (2×10^{15} cm^{-2} at 75 kV). A heavily doped p-type substrate contact was also defined using boron implantation (1×10^{15} cm^{-2} at 50 kV). E-beam lithography and RIE was used to define the SET (50 nm-wide and 1 µm-long nanowire) and the memory node (14 µm × 24 µm rectangular region).

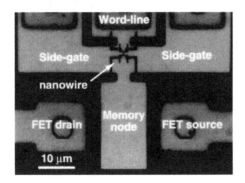

Fig. 4.16 Optical micrograph of a single-gate L-SEM cell. (Reprinted with permission from Durrani *et al.*, *Appl. Phys. Lett.* **74**, 1293 [1999]. Copyright 1999, American Institute of Physics).

A ~20 nm thick oxide layer was then grown to passivate and thin the top silicon. Contacts to the cell were fabricated using a CMOS-type 'back-end' metallization process, as follows. First, a 500 nm-thick SiO$_2$ dielectric was sputter-deposited, and subsequently H$_2$-passivated by annealing for 15 minutes at 400 °C in an ambient of forming gas (5% H$_2$ in N$_2$) at atmospheric pressure. Sloping-profile contact holes were wet etched selectively through the SiO$_2$ dielectric, and a Ti/Al$_{0.99}$Si$_{0.01}$/Ti blanket metallization (80 nm/500 nm/80nm) was sputter-deposited. The excess metal was removed by SiCl$_4$ RIE, after which the contacts were sintered for 15 minutes at 200 °C in forming gas. Figure 4.16 shows an optical micrograph of the cell. The SET and MOSFET were biased using a word-line voltage V_{ws} and drain-source voltage V_{ds}, respectively. Applying a voltage V_{trim} on the side-gates varied the SET Coulomb gap.

Figure 4.17 shows the drain-source current vs. gate voltage (I_{ds}-V_{gs}) characteristics at 4.2 K, in a MOSFET without a coupled SET. The top-right and bottom-right insets show the corresponding drain-source current-voltage (I_{ds}-V_{ds}) characteristics and the I_{ds}-V_{gs} characteristics on a log-linear scale. The peak transconductance for this device was 14 mS/mm.

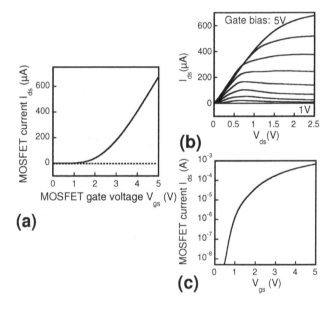

Fig. 4.17 L-SEM: Electrical characteristics of the sense MOSFET. (a) I_{ds}-V_{gs} characteristics on a linear scale. (b) I_{ds}-V_{ds} characteristics. (c) I_{ds}-V_{gs} characteristics on a log-linear scale.

The hysteresis characteristics of the L-SEM memory cell at 4.2 K are shown in Fig. 4.18(a). As the word-line voltage V_{ws} is swept cyclically, a stable hysteresis is observed in the MOSFET drain-source current I_{ds}. The hysteresis can be observed for sweep times in excess of one hour. The SET side-gate voltage V_{trim} strongly modulates the hysteresis. We explain the characteristics with reference to the circuit of Fig. 4.18(b), assuming, for simplicity, a fixed, symmetric Coulomb gap in the SET of $2V_C$. With V_{ws} initially at zero, the voltage across the SET (V_{SET}) is less than the Coulomb gap, and the SET current (I_{SET}) is zero. As V_{ws} increases and V_{SET} exceeds the positive edge of the Coulomb gap $+V_C$, charge (positive) is transferred onto the memory node, and the memory-node voltage V_m increases with the SET voltage maintained at $+V_C$. The increase in V_m is detected by the MOSFET, and there is an increase in I_{ds}. If the direction of the V_{ws} sweep is now reversed, the SET voltage immediately falls below $+V_C$, and it is not possible for the memory node to discharge. I_{ds} therefore remains constant.

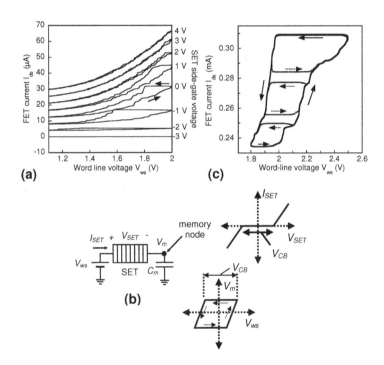

Fig. 4.18 Single-gate L-SEM: Hysteresis characteristics at 4.2 K. (a) Hysteresis in sense MOSFET current. (b) L-SEM operation. (c) 'Universal' hysteresis.

When the SET voltage falls to $-V_C$, the memory node begins to discharge and I_{ds} falls. There are two different memory states, corresponding to voltages separated by the total SET gap of $2V_C$, and this leads to a hysteresis in the MOSFET characteristics. Experimentally, the width of the hysteresis is not constant because the SET Coulomb gap is not invariant. This is discussed in more detail below.

The hysteresis is strongly influenced by the SET side-gate voltage V_{trim} (Fig. 4.18[b]). In the characteristics of Fig. 4.18(a), when V_{trim} = -3 V, conduction in the nanowire of the SET is pinched off, isolating the memory node from the word-line, and I_{ds} is nearly constant. At the other extreme, V_{trim} = +4 V, nanowire conduction is enhanced to an extent that a Coulomb gap does not exist. The SET then simply acts as a resistor in series with the memory node, and the memory cell characteristics are not hysteretic. At all intermediate values of V_{trim}, the Coulomb gap in the SET separates V_m and V_{ws}, and a hysteresis is seen. Figure 4.18(c) shows the hysteresis in a memory cell at V_{trim} = 0 V and V_{ds} = 4 V, as V_{ws} is swept between 1.8 V and 2.6 V. Four different sweeps over different voltage ranges demonstrate that the legs of the hysteresis are stable.

An interesting feature in Fig. 4.18(c) is the observation of clear steps in I_{ds}. These steps are associated with the single-electron oscillations in the SET. Although the applied SET side-gate voltage is fixed, there is an increase in the voltage at either end of the nanowire, i.e. V_{ws} and V_m. This implies that the voltage between the side-gate and the nanowire *decreases* as V_{ws} is swept up. This change in the side-gate-to-nanowire voltage causes single-electron oscillations in the SET, accompanied by a modulation of the width of the Coulomb gap, in a manner qualitatively similar to the characteristics in Fig. 4.12. If the Coulomb gap increases, the SET moves into the zero-current regime and V_m remains constant until the increased Coulomb gap is overcome at higher V_{ws}. This leads to a step in I_{ds}. We also note that, with increasing V_{ws}, the voltage between the side-gate and the nanowire ultimately falls to an extent that the SET is fully pinched off, leading to saturation of the MOSFET current I_{ds}, as seen in the hysteresis at V_{trim} = –1 V in Fig. 4.18(a).

Our experimental single-gate L-SEM cell has a very large memory node 14 µm × 24 µm in area, with a gate capacitance of ~10^{-13} F. The two memory states then correspond to a difference of around 10^6 electrons on the memory node. However, in this design the memory node is easily scalable to much smaller sizes and can potentially operate with very few electrons. With scaling of MOSFETs into the sub-50 nm gate width regime, the single-gate L-SEM cell becomes a much more promising design.

4.5.3 Split-gate L-SEM

The L-SEM memory node size, and therefore the stored charge, may be greatly reduced using a split-gate L-SEM cell. This cell uses a MOSFET with dual, in-plane top-gates (Fig. 4.19). Here, two 'outer gate' regions are used to form the MOSFET channel on either side of a central, reduced area memory node (or 'central gate'). The MOSFET current is controlled by the gating effect of the memory node, while sensing the cell. The separation between the outer gates and the memory node must be sufficiently small to ensure adequate continuity of the channel from drain to source. The outer gates can also be employed to close the channel, allowing the cell to be deselected in a memory array application.

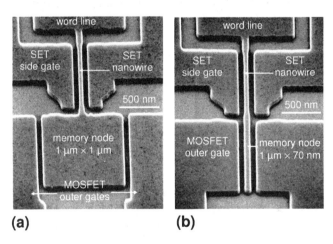

Fig. 4.19 (a) Split-gate L-SEM with a large-area memory node. (b) Split-gate L-SEM cell with a scaled memory node.

The split-gate cell is fabricated in a manner similar to the single-gate L-SEM cell, in wafer-bonded SOI material. The basic cell structure is only modified by the addition of the split-gate, defined using e-beam lithography and reactive-ion etching. The separation between the memory node and the adjacent outer gates is ~70 nm. Figure 4.19 shows scanning-electron micrographs of two different split-gate L-SEM cells prior to growth of the passivation oxide. In Fig. 4.19(a), a cell with a 1 μm × 1 μm area memory node is shown, and in Fig. 4.19(b), a cell with a 1 μm × 70 nm memory node is shown. From 2-D capacitance calculations we estimate that, for a typical SET Coulomb gap of 0.1 V, the memory states in the device of Fig. 4.19(a) differ by ~600 electrons, and the memory states in the device of Fig. 4.19(b) differ by ~100 electrons. These figures are considerably lower than the ~50,000 electrons used in standard, classical DRAM devices.

The operation of the MOSFET outer gates is demonstrated in Fig. 4.20, where the transfer characteristic using the central gate is shown at different outer-gate voltages V_{outer}. The central gate (also the memory node) of this MOSFET was 1 μm × 1 μm in area and the SET side-gate voltage was 5 V. Under this biasing condition, the SET is effectively a simple resistor and the central MOSFET gate voltage V_m is only insignificantly different from the word-line voltage V_{ws}. We note that, for V_{outer} = 1 V (not shown), I_{ds} < 10^{-11} A.

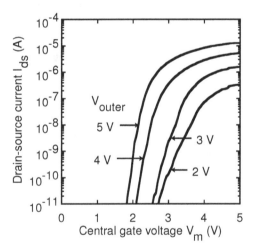

Fig. 4.20 Split-gate sense MOSFET *I-V* characteristics.

For a fixed central-gate voltage, V_{outer} can be used to control I_{ds} and switch the MOSFET 'on' or 'off'. In a memory array application, it is therefore possible to select the split-gate memory cell by controlling the outer gates of the MOSFET. However, the capacitive coupling between the central gate and the outer gates can, in principle, produce an unwanted change in the memory state. This limits the useable range of voltages applied to the outer gate.

Figure 4.21 shows the hysteresis characteristics of the split-gate L-SEM at 4.2 K, for two cells with memory node dimensions (a) 1 μm × 1 μm and (b) 1 μm × 70 nm. The MOSFET current I_{ds} is shown at fixed V_{ds} as V_{ws} is swept cyclically. The SET side-gate voltage V_{trim} can be used to control the width of the hysteresis. The MOSFET outer-gate voltage V_{outer} is fixed at 5 V to ensure that there is a continuous, open drain-source channel in the MOSFET. The behaviour of the cell is qualitatively similar to the single-gate L-SEM cell, where the MOSFET current exhibits a clear hysteresis, strongly modulated by the SET side-gate voltage. As before, the separation between the sweep-up and sweep-down curves depends on the Coulomb gap in the SET for each specific biasing condition.

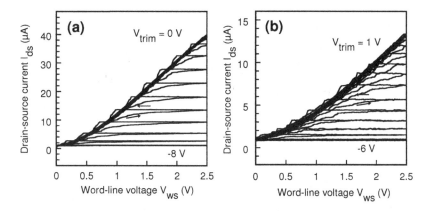

Fig. 4.21 Split-gate L-SEM cell characteristics at 4.2 K, for cells with memory node dimensions. (a) 1 μm × 1 μm. (b) 1 μm × 70 nm.

The speed of operation of the L-SEM is characterized in Fig. 4.22(a). Here, a cell with a 1 μm × 1 μm memory node is operated for a range of write pulse widths from 1 μs to 10 ns. In this measurement, the word-line is held at a steady state word-line voltage V_{wss}, with a series of superimposed, alternating 'write-0' and 'write-1' pulses (Fig. 4.22[b]). The voltage level and duration of these pulses are $+V_1$ and $-V_0$, and t_1 and t_0, respectively. V_{ds} is held at a constant 5 V. The write pulse width is shorter than the MOSFET current sampling interval, and the plots represent the MOSFET current I_{ds} during successive 'read' intervals 25 ms in duration.

The cell operation is essentially similar to a hysteresis sweep in the interval $V_{wss} - V_0 < V_{wss} < V_{wss} + V_1$, (Fig. 4.22[c]), shown for three values of V_{trim}). Consequently, for any given intermediate voltage V_{wss}, there are two possible values of the cell output current I_{ds}. In the characteristics of Fig. 4.22(a), t_0 is fixed at 1 μs and t_1 is reduced. It is seen that, even at $t_1 = 10$ ns, the memory-node charging rate is sufficient for two values of I_{ds} to be resolved. This performance is consistent with a simple estimate of a time constant, ~0.1 ns, for a memory-node capacitance ~1 fF and a nanowire resistance ~100 kΩ.

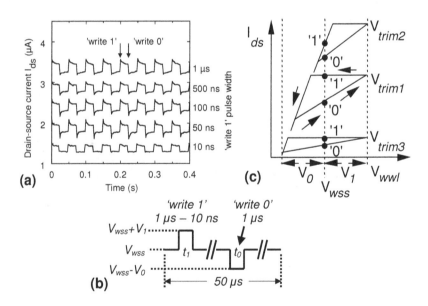

Fig. 4.22 Pulsed operation of the L-SEM cell. (a) Pulsed characteristics for 'write' pulse widths from 1 μs to 10 ns. The curves are offset from each other by 0.5 μA, for clarity. (b) Schematic diagram of pulses. (c) Schematic diagram of hysteresis.

4.5.4 L-SEM 3×3 cell array

We have used the split-gate L-SEM cell in crystalline SOI material to fabricate a 3 × 3 bit experimental memory circuit (Fig. 4.23). In this circuit, the selection and read/write process for a cell within the array has been demonstrated and the effect of the operating temperature on cell performance characterized. Arrays were fabricated using the 1 µm × 1 µm memory node, and the 1 µm × 70 nm memory node, split-gate L-SEM. The arrays used two layers of interconnects to form the 'back-end'. Here, the first layer of interconnects was fabricated in the top silicon of the SOI material, using RIE for isolation. The second layer of interconnects was fabricated using the metallization scheme already developed for the individual single-gate and split-gate memory cells, where Ti/Al$_{0.99}$Si$_{0.01}$/Ti layers were supported on a sputter-deposited SiO$_2$ dielectric. At interconnect crossovers and at the cell contacts, wet etched holes in the dielectric were used. Figure 4.23(a) shows an optical micrograph of the final 3 × 3 cell array.

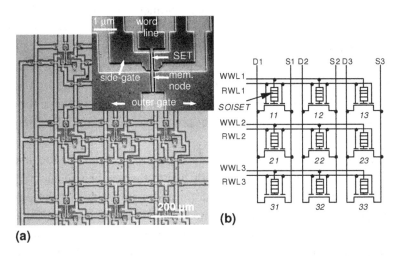

Fig. 4.23 Split-gate L-SEM 3 × 3 array. (a) Optical micrograph of array. The inset shows a scanning electron micrograph of a cell. (b) Circuit diagram.

The memory circuit design (Fig. 4.23[b]) was developed by Hitachi Cambridge Laboratory, University of Cambridge. In each row of cells, the SET is connected to a write word-line (WWL1, WWL2 or WWL3), used for cell selection during writing. Similarly, the outer gates of the split-gate MOSFET are connected to a read word-line (RWL1, RWL2 or RWL3), used for cell selection during read-out. The split-gate MOSFETs are connected between a pair of drain and source lines arranged as columns (D1-S1, D2-S2 and D3-S3), with three cells along each column. These lines are used to transfer data into the cells and to sense a current from a selected cell. The side-gates of the SET are connected to a common voltage input, the 'trimming gate', with an applied voltage V_{trim}.

4.5.4.1 Memory cell selection

We now discuss the pulsed operation of a single selected memory cell, e.g. cell '11' within the 3×3 array. Figure 4.24 shows the control pulses and output from a 1 µm × 1 µm memory node cell, measured at 20 K. Voltage pulses V_{wwl1} are applied to the write word-line 'WWL1'. To write to the cell, voltage pulses V_{d1} and V_{s1} are applied to the corresponding drain (D1) and source (S1) lines. At the point marked 'A', a –0.4 V pulse immediately followed by a +0.4 V pulse is superimposed on the steady state, write word-line voltage $V_{wwl1} = V_{wss1} = 2.18$ V. Together, V_{wwl} and V_{trim} bias the cell into a regime where a memory hysteresis exists (Fig. 4.22). A steady state read word voltage $V_{rwl1} = V_{rss1} = 3.6$ V, applied to 'RWL1', biases the channel region under the split gate, just below the threshold voltage, such that the MOSFET current I_{ds1} remains low.

The ±0.4 V write word-line pulses at 'A' are chosen to be too small to overcome the SET Coulomb blockade and to thereby change the memory state of the cell. However, if the drain and source lines of the cell are pulsed simultaneously with a common 'data' voltage, then the Coulomb blockade is overcome and a '1' or '0' is written. We write the data bit '1' into the cell by pulsing both V_{d1} and V_{s1} 'low' to –0.22 V (marked 'B') simultaneously with the ±0.4 V write word-line pulses. This makes the voltage across the SET, which is proportional to $V_{wwl1} - V_{d1}$ (or $V_{wwl1} - V_{s1}$), large enough to overcome the Coulomb blockade during the +0.4 V pulse at 'A', transferring electrons from the memory node to the write word-line. This writes a '1' into the cell. Similarly, a '0' is written to the cell at *'C'*, when the +0.22 V drain and source line pulses overlap with the –0.4 V write word-line pulse. Only cell '11' is accessed because the drain, source and write word-line pulses must be applied simultaneously in order to write to a cell. The other cells along the column 'D1', 'S1' do not receive write word-line pulses and the other cells along the row 'WWL1' do not receive drain and source line pulses.

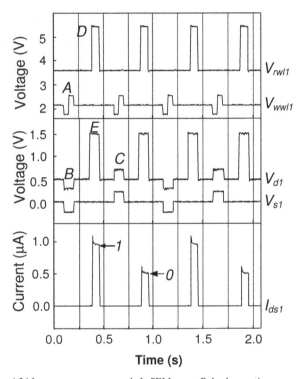

Fig. 4.24 Large-area memory-node L-SEM array: Pulsed operation.

The '1' written by the 'data' pulses at *B* is sensed at a high current level in I_{ds1} by pulsing V_{rwl1} to enable the cell (marked *'D'*) and pulsing V_{d1} (marked *'E'*) to generate a current. The '0' written by the 'data' pulses at *'C'* is detected in a similar manner, at a lower level of I_{ds1}. Other cells along 'RWL1' have zero drain-source voltage and do not generate a sense current, while other cells along the column 'D1', 'S1' lack a read pulse and are not selected. We intentionally pulse the drain line 0.02 s earlier than the read word-line in order to verify that a current is measured only when the pulses overlap. The basic write/read cycle for the memory states is repeated further and no attempt has been made to reduce the cycle time. The current difference between the two memory states is 0.5 µA or 50% of the '1' state current level. This can be used, with the MOSFET transconductance of ~3 µAV^{-1}µm^{-1}, to estimate the difference in the memory-node voltage for the two states. The memory-node states are separated by ~0.17 V. In this device, using 2-D memory-node capacitance calculation, we estimate that ~6,000 electrons are added to the memory node per volt change in the memory-node voltage. This implies that the '0' and '1' states differ by ~1,000 electrons on the memory node.

The amplitude of the drain and read word-line pulses are chosen such that there is an appreciable current swing in I_{ds1}. However, as both the drain region and the outer gate couple capacitively to the memory node, large pulse levels may overcome the SET Coulomb gap and disturb the memory states. In our device, the memory-node-to-drain capacitance and the memory-node-to-outer-gate capacitance are smaller than the memory-node-to-channel capacitance. This implies that one can apply drain and read-line voltages relatively larger in magnitude than the voltage level in an intentional write operation ($V_{wwl1} - V_{d1} = V_{wwl1} - V_{s1}$) without destroying the memory states.

Figure 4.25 shows the operation of an array using split-gate cells with 1 µm × 70 nm memory nodes. The cell selection scheme is similar to that shown in Fig. 4.24. We also explicitly verify that both write word-line and drain/source line pulses are necessary to write to the cell and that the cell contents are not disturbed if information is written to an adjacent cell. The measurement uses cells '31' and '32', connected along the same row ('WWL3', 'RWL3'), but in different columns ('D1', 'S1' and 'D2', 'S2'). Basic memory operation is demonstrated in cell '32', using the pulses V_{wwl3}, V_{d2} and V_{s2} on lines 'WWL3', 'D2' and 'S2', respectively, to write a '1' and a '0'. A voltage V_{rwl3} on 'RWL3' is used to read out the cell current I_{ds2}. The difference in I_{ds2} between the two states is 0.25 µA or 25% of the '1' state. This corresponds to a difference of 0.06 V in the memory-node voltage (the MOSFET transconductance was 4.2 µAV^{-1}µm^{-1}). Using a figure of ~1,000 electrons per volt change in memory-node voltage in the cells, we estimate that the '0' and '1' states differ by ~60 electrons on the memory node. If we use drain and source line pulses V_{d20} and V_{s20} where the write '1' pulses (marked '*A*') have been removed, only the '0' state is written and sensed (I_{ds20}). This demonstrates that both write word-line and drain/source line pulses must be applied in order to write to the cell.

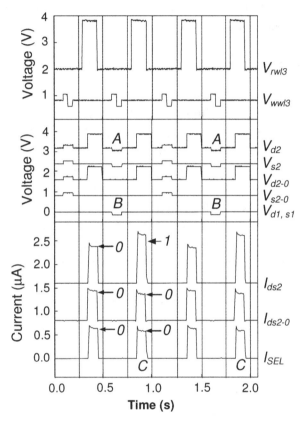

Fig. 4.25 Scaled memory-node L-SEM array: Pulsed operation.

We now demonstrate that a '0', stored in cell '32', is undisturbed by the writing of the opposite state '1' to cell '31'. A '1' is written to cell '31' by applying the write '1' pulses V_{d1}, V_{s1} to 'D1', 'S1' (marked '*B*'). It is seen that when the output of cell '32' is sensed again (I_{SEL}, at '*C*'), it still retains the '0' written earlier. In this measurement, V_{trim} was chosen such that a memory effect was observed independently in both cells for identical bias and pulse voltages. The two cells displayed optimum performance at slightly different values of V_{trim}, implying that the memory state separation was not maximized in either cell. For example, if cell '31' was re-measured with $V_{trim} = -1.9$ V, a 30% difference in I_{ds} between the '1' and '0' states was observed. It may be possible to reduce the spread in the electrical characteristics of these cells by optimizing the SET design with a view to obtaining the same Coulomb gaps at the same value of side-gate voltage.

4.5.4.2 Temperature dependence of memory cell characteristics

As the operating temperature of an SET is increased, thermally activated current begins to flow within the Coulomb gap. Because the L-SEM uses the Coulomb gap to trap charge on the memory node, a non-zero sub-gap current leads to a fall in the memory state retention time. The retention time depends on a time constant, determined by the memory-node capacitance and the non-linear resistance of the SET. As long as the SET I-V characteristics remain non-linear and a significantly higher resistance is observed within the Coulomb gap, charge can be stored on the memory node for longer than it takes to transfer charge to the memory node. This is because the charge transfer operation is performed with the SET biased well outside the Coulomb gap, and the charging rate for this process is lower. If the temperature increases to an extent that the retention time is comparable to the write time, we may infer that the Coulomb gap has disappeared completely, the SET I-V characteristics are linear, and memory operation has ceased.

The temperature-dependence of the memory effect in a 1 μm × 70 nm memory-node cell is shown in Fig. 4.26. In Fig. 4.26(a), we observe that the memory states (continuously switched, in a manner similar to Fig. 4.22[a]) begin to discharge as the temperature increases from 20 K to 50 K, and the retention time deteriorates. This can be associated with an increasing sub-gap current in the SET. In Fig. 4.26(b), we plot the temperature-dependence of the time taken by the '1' state to decay by 50%. The data are obtained from a measurement similar to Fig. 4.26(a), with 50 ms write '1' and write '0' pulses. The rise time of these pulses was 0.2 ms. There was no measurable lag of the MOSFET output current with the write pulse rise time. The data show that there is an approximately logarithmic fall in the retention time with temperature, and above 65 K, the retention time is barely longer than the pulse rise time, i.e. single-electron controlled operation has disappeared. Improvement in the temperature of operation of the SET would directly improve the cell operating temperature. In this regard, the nanocrystalline silicon 'point-contact' SET (Tan et al., 2003), capable of operating at room temperature, is of great potential for the fabrication of an L-SEM operating at room temperature.

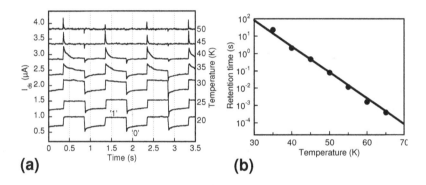

Fig. 4.26 (a) Temperature-dependence of the pulsed electrical characteristics of a split-gate L-SEM cell. (b) Retention time vs. temperature.

Finally, we comment briefly on the reproducibility of the L-SEM characteristics. Good reproducibility of the L-SEM characteristics requires well-matched, reproducible SET characteristics. Regarding the reproducibility of crystalline silicon nanowire SETs, in nine different devices using relatively thin wires of ~30 nm width, a mean Coulomb gap of 0.3 V and a standard deviation of 0.09 V was measured. Similarly, in the 3×3 L-SEM cell array, a memory effect could be observed in all the cells within a trimming-gate voltage range of 1 V. We note that the L-SEM devices in this work were fabricated in a laboratory, and for a detailed investigation of the reproducibility of these devices, it would be necessary to optimize the various fabrication processes in clean room conditions.

Chapter 5

Few-Electron Transfer Devices

5.1 Introduction

Single-electron charging effects can be used as the basis of devices where it is possible to control the transfer of charge packets formed by single, or only a few, electrons using one or more AC signals. The most widely investigated of these devices are 'single-electron turnstiles' (Geerligs et al., 1990) and 'single-electron pumps' (Pothier et al., 1991), where the transfer of electrons is controlled by a radio-frequency (r.f.) voltage signal. A single electron, or a small number of electrons, can be transferred through these devices in each cycle of the r.f. signal. The device current is then given by the simple relation $I = nef$, where f is the frequency of the r.f. signal and n is an integer. Other few-electron transfer devices are also possible, e.g. devices analogous to charge-coupled devices (CCDs) (Fujiwara et al., 2001), devices where few-electron packets are controlled using multiple-tunnel junction (MTJ) SETs (Tsukagoshi et al., 1997) or devices using integrated nanoscale MOSFETs and SETs (Ono and Takahashi, 2003).

Early investigations of single-electron turnstiles and pumps were driven by the possibility of metrological applications, e.g. for the fabrication of a fundamental standard for current and capacitance (Odintsov, 1991; Martinis et al., 1994). More recently, with great improvements in the performance of SETs and single-electron memory, there is additional interest in few-electron transfer devices as a means to control the flow of few-electron 'bits' between various parts of a nanoscale-integrated circuit. It may also be possible to realize single-electron logic circuits using these devices as basic components, by means of the 'binary decision diagram' architecture proposed by Asahi *et al.* (Asahi *et al.*, 1997; Chapter 6).

This chapter discusses in detail the design, fabrication and operation of single-electron transfer circuits. The earliest devices, the single-electron turnstile and pump, are considered first (Section 5.2). We note that, while these devices have been mostly implemented in the Al/AlO$_x$ and the GaAs/AlGaAs 2-DEG system and not in silicon, the design of other electron transfer systems rely on the concepts developed here. We then consider bi-directional electron pumps using MTJs. These devices, fabricated initially in δ-doped GaAs but then investigated extensively in silicon-on-insulator (SOI) material (Section 5.3), may be used to transfer few-electron packets via one or more r.f. signals. Section 5.4 considers devices in silicon capable of one electron transfer, e.g. single-electron pumps using nanoscale CCDs and SET/MOSFET hybrid electron pumps and turnstiles. Finally, in Section 5.5 we introduce metrological applications of single-electron transfer circuits.

5.2 Single-Electron Turnstiles and Pumps

At the start of the 1990s, three different types of single-electron transfer devices were demonstrated within a short period. These devices were: (i) the single-electron turnstile of Geerligs *et al.* (Geerligs *et al.*, 1990), based on four metal islands and operated using a single AC signal, (ii) the single-electron pump of Pothier *et al.* (Pothier *et al.*, 1991), based on two metal islands and operated using two phase-locked AC signals and (iii) the combined single-electron turnstile and pump of Kouwenhoven *et al.* (Kouwenhoven *et al.*, 1991b, 1991c), based on a semiconductor quantum dot and operated using two phase-locked AC signals. The concepts developed by these devices are used in almost all other implementations of few-electron transfer devices. In the following, we discuss all three devices in detail.

5.2.1 Single-electron turnstile

Within a few years of Fulton and Dolan's (Fulton and Dolan, 1987) demonstration of the first SET using Al/Al$_2$O$_3$ tunnel junctions, the first single-electron turnstile circuit was demonstrated by Geerligs *et al.* (Geerligs *et al.*, 1990). This device uses an r.f. gate voltage to transfer one electron per r.f. cycle through a series of tunnel junctions, in a manner analogous to a shift register operating with single electrons. We now consider the operation of the device in detail.

Figure 5.1(a) shows the circuit diagram of the device. The device consists of four tunnel junctions, with capacitance and resistance C_m and R_m, respectively, for the mth junction. The central island is biased using a gate capacitance C_g by an AC gate voltage v_g, of frequency f. A DC bias of $V/2$ and $-V/2$ is applied to either end of the circuit, asymmetrically biasing the circuit.

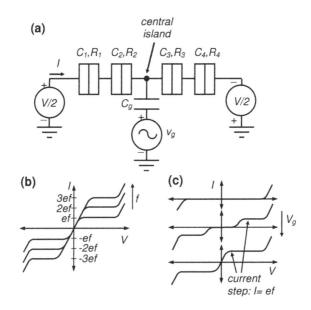

Fig. 5.1 Single-electron turnstile. (a) Circuit diagram. (b) Schematic *I-V* characteristics vs. r.f. frequency f. (c) Schematic *I-V* characteristics vs. gate voltage V_g (after Geerligs *et al.*, 1990).

It is possible to increase the number of tunnel junctions in the arms on either side of the central island to greater than two, as long as the symmetrical T-shape of the circuit is maintained. The device requires at least two tunnel junctions on either side to prevent unwanted tunnelling of charge across the device. It is also useful for the device to include two additional gates (not shown in Fig. 5.1[a]), capacitively coupled to the islands on either side of the main island. These gates can be used to tune out the effect of background 'offset' charges on the islands, and to obtain symmetrical Coulomb blockade regions in both arms of the circuit. Geerligs et al. fabricated this device using 0.5 fF and 340 kΩ Al/Al$_2$O$_3$ tunnel junctions, and a central gate capacitance of 0.3 fF.

The circuit operates as follows. An electron can tunnel across a specific tunnel junction m in the circuit if the applied biases V and v_g are such that the junction voltage overcomes the single-electron charging energy of the junction. This occurs when the absolute value of charge across the junction $|Q|$ exceeds a critical charge Q_{cm} on the junction (see Chapter 2 for a discussion of the 'critical charge' in single-electron tunnelling). This is given by $Q_{cm} = e/2(1 + C_{em}/C_m)$, where C_m is the junction capacitance and C_{em} is the equivalent capacitance of the rest of the circuit, lying in parallel with C_m. Geerligs et al. considered a specific condition, where all the tunnel junction capacitances are equal, $C_m = C$, and the gate capacitance $C_g = C/2$. For this condition, it can be shown that $Q_{cm} = Q_c = e/3$ for all the tunnel junctions. Now, as v_g increases in a cycle from zero to a large enough positive value, Q_c is exceeded in the tunnel junctions in the left arm of the circuit, but not for those in the right arm, due to the polarity of the applied DC bias V. A single electron then tunnels across the left arm onto the central island. This electron polarizes C_g, reducing the charge across each junction in the left arm to below Q_c, so that it is not possible for another electron to tunnel onto the island. One electron is then trapped on the central island. Later in the cycle, as v_g decreases to negative values, Q_c is exceeded for the tunnel junctions in the right arm and the electron leaves the central island. The net result is the transfer of one electron through the turnstile in each cycle of v_g, such that the device current is given by $I = ef$.

Geerligs et al. operated their turnstiles in a dilution refrigerator at frequencies from 4 to 30 MHz. The device temperature, estimated by comparison with simulation, was 50–75 mK. A magnetic field of 2 T was applied such that the device operated in the 'normal' non-superconducting state. Figure 5.1(b) and Fig. 5.1(c) schematically show the device I-V characteristics, as the frequency f and the amplitude (V_g) of v_g, respectively, are varied. As a function of frequency, the current shows plateaus corresponding to $I = nef$. These plateaus are associated with one electron transferred per cycle through the turnstile. As a function of the amplitude of v_g, if this is zero, i.e. no AC gate voltage is applied, a large Coulomb gap is observed in the I-V characteristics. This is created by the MTJ formed by the four tunnel junctions. As the amplitude of v_g is increased, turnstile action begins to occur with the formation of a current plateau, initially at values of V above a threshold voltage, and then near $V = 0$. In the device of Geerligs et al., the current plateaus were flat to within the noise level up to a frequency of 10 MHz. However, from 10 to 30 MHz, an increasing deviation was observed, and for 30 MHz, the error in I was ~10^{-2} of the average value of I.

The accuracy of the turnstile in following $I = ef$ depends on factors which are different at low or high frequencies. At low frequencies, it is possible for an electron trapped within the turnstile to escape, either by thermal excitation, or by a higher order tunnelling process (Jensen and Martinis, 1992). Clearly, for lower measurement temperatures, lower operating frequencies can be used for a required accuracy. At higher frequencies, the accuracy depends on the probability that tunnelling of an electron can take place within the time period of the AC signal, i.e. $f \gg 1/(RC)$. Regarding the low frequency limit, it is necessary to compare the tunnelling rate Γ_1 for unwanted tunnelling events to the rate Γ_2 for wanted tunnelling events. The ratio of Γ_1 and Γ_2 can be shown to be ~$\exp(-\Delta E/k_B T)$, where ΔE is the change in electrostatic energy for a tunnelling event. For the device of Geerligs et al., $\Delta E \sim 0.1 e^2/2C$, which implied that the measurement temperature T had to be small enough for $k_B T \ll 0.1 e^2/2C$. For C ~0.1 fF, this gave a very low operating temperature of $T < 100$ mK. Considerable improvement in this is possible with smaller values of C. Regarding the high frequency limit, we need to compare the tunnelling rate $\Gamma \sim 1/(RC)$ to the frequency f. For a tunnelling rate far higher than the frequency, e.g. $\Gamma = 1000f$, to make sure that an electron can tunnel, we obtain a limit on the maximum frequency, $f < 10^{-3}/RC$. For $R \sim 300$ kΩ and $C \sim 0.1$ fF, this given $f < 33$ MHz.

5.2.2 Single-electron pump

We now consider the single-electron pump of Pothier *et al.* (Pothier *et al.*, 1991, 1992), where electrons can be transferred by AC signals *against* the effect of an applied DC bias, or transferred without the application of a DC bias. In contrast, the single-electron turnstile of Geerligs *et al.* requires both an AC signal and a DC bias to drive current through the device (Fig. 5.1). Due to the symmetry of the two arms of the turnstile, without a DC bias the gate voltage overcomes the charging energy of the tunnel junctions in both arms simultaneously and it is not possible to observe a net current flow through the device.

The circuit diagram of the single-electron pump of Pothier *et al.* is shown in Fig. 5.2(a). The device may be realized using three Al/Al$_2$O$_3$ tunnel junctions with two Al islands, tunnel-coupled to each other. In addition, the islands are coupled capacitively to two AC gate voltages V_1 and V_2, with a constant phase difference. Finally, the right-hand side island is tunnel-coupled to ground and the left-hand side island is coupled to a DC voltage source V. This DC voltage is not essential for circuit operation – the pump can operate at $V = 0$ – and is present only to characterize the pumped current against the effect of this. We note that the circuit configuration is identical to two coupled quantum dots (Chapter 3), but with AC rather than DC gate voltages.

For small values of V well inside the Coulomb gap, the electron numbers on the two islands, n_1 and n_2, respectively, are determined by the values of the gate voltages V_1 and V_2. Due to the central tunnel junction capacitance C_2, both V_1 and V_2 couple capacitively to Island 1. Similarly, V_1 and V_2 also couple capacitively to Island 2. In addition, electrons trapped on Island 1 electrostatically bias Island 2, and vice versa. These effects lead to the charge stability diagram shown in Fig. 5.2(b). Here, the charge configuration (n_1, n_2) of the device, as a function of V_1 and V_2, is stable within hexagonal-shaped regions. Crossing the boundary of a hexagonal region leads to a change in either n_1 or n_2 by one. A triple point is formed at the corners of the hexagonal regions, with three stable charge configurations in close proximity, e.g. the triple point lying within the dotted circle is surrounded by the configurations (0,0), (1,0) and (1,1). A measurement of the current I through the circuit as a function of DC values of V_1 and V_2, with a small, constant bias V, leads to a series of current peaks at the triple points. Current flows only at these points because it is only here that the Coulomb blockade of both islands is simultaneously overcome.

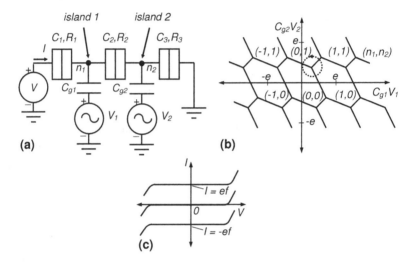

Fig. 5.2 Single-electron pump. (a) Circuit diagram. (b). Stability diagram. (c) Schematic I-V characteristics with r.f. frequency f (after Pothier *et al.*, 1991, 1992).

Now, if the values of V_1 and V_2 are changed such that a circular path is followed around a triple point, i.e. the path along the dotted circle, then an electron is transferred across tunnel junction T1 onto Island 1 (configuration [0,0] → [1,0]), then from Island 1 onto Island 2 (configuration [1,0] → [0,1]) and finally across tunnel junction T3 onto the right electrode (configuration [0,1] → [0,0]). It is possible to change V_1 and V_2 in this manner using synchronized AC signals with a given phase difference. One electron is pumped through the circuit in each cycle of V_1 and V_2, and the direction of the current is given only by the sense of rotation around the triple point.

Fig. 5.2(c) schematically shows the *I-V* characteristics of the device. For zero V_1 and V_2, a wide Coulomb gap corresponding to the total charging energy of the three tunnel junctions is observed. With V_1 and V_2 applied, a pumped current $I = \pm ef$ is observed within the Coulomb gap, where the sign is a function of the phase difference. *I* has a non-zero value at $V = 0$ V and even for negative values of *V*, demonstrating that electrons can be pumped through the device even against an applied bias. Pothier *et al.* (Pothier *et al.*, 1991, 1992) operated their pump for a frequency range of 2–20 MHz, at milli-Kelvin temperatures. They estimated that the tunnel capacitances $C_1 = C_2 = C_3 = 1.5$ fF, and the gate capacitances $C_{g1} = C_{g2} = 0.02$ fF.

5.2.3 Single-electron turnstile and pump using a semiconductor quantum dot

Kouwenhoven *et al.* (Kouwenhoven *et al.*, 1991b, 1991c) demonstrated that a semiconductor quantum dot can be operated as both a single-electron turnstile and pump, using two r.f. signals to modulate the tunnel barriers isolating the quantum dot. The device was also proposed, and theoretically analysed, by Odintsov (Odintsov, 1991). Kouwenhoven *et al.* fabricated the device using a lateral quantum dot defined in the 2-DEG formed by an AlGaAs/GaAs heterostructure (Shur, 1987). Such a quantum dot can be defined using planar metal gates on the surface of the sample to deplete regions of the 2-DEG, forming isolation regions and tunnel barriers (Fig. 5.3). Alternatively, a combination of mesa etching and surface gates may be used to define the quantum dot (Nagamune *et al.*, 1994).

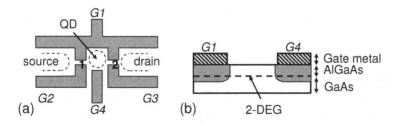

Fig. 5.3 Single-electron turnstile and pump using a semiconductor quantum dot.

Figure 5.3 illustrates the former, more commonly used scheme in plan and cross-sectional view. Here, the gates G1, G2, G3 and G4 are biased using a negative voltage such that the underlying 2-DEG is depleted, defining a quantum dot in the central region. In the device of Kouwenhoven *et al.* (Kouwenhoven *et al.*, 1991b, 1991c), the quantum dot was ~0.8 μm in diameter. Gates G2–G4 are biased such that the 2-DEG in the space between them is depleted, while G1 is biased such that a tunnel barrier is formed at the constrictions 1 and 2. The quantum dot is then tunnel-coupled across the potential barriers at 1 and 2 to 2-DEG source and drain regions.

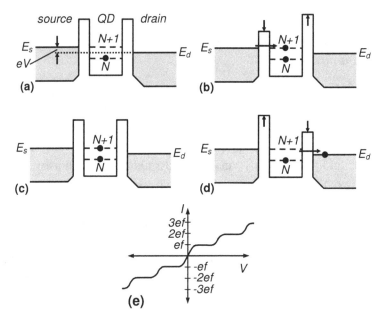

Fig. 5.4 Operation of the quantum dot single-electron turnstile and pump (after Kouwenhoven *et al.*, 1991b, 1991c).

We explain the operation of the device using the energy diagram of the device, shown in Fig. 5.4. The Fermi energy of the source and drain regions is E_s and E_d respectively. Quantum confinement of electrons by the tunnel barriers leads to N discrete energy levels with energy E_N in the quantum dot, and the energy difference between the Nth and $N+1^{th}$ level is given by $\Delta E = E_{N+1} - E_N + e^2/C$, where C is the capacitance of the quantum dot. For the device of Kouwenhoven et al., $E_{N+1} - E_N$ was small compared to e^2/C and could be ignored.

Figure 5.4(a) shows the device without the application of AC voltages. When a small bias V is applied between source and drain, electrons can tunnel through the quantum dot if an energy level, e.g. the level $N+1$, lies between E_s and E_d. Electron tunnelling, and the current, is reduced to zero when the quantum dot is in Coulomb blockade. This occurs when the voltage on G4 (V_{g4}) is made more negative, raising the bottom of the conduction band in the quantum dot such that E_{N+1} lies above E_s, but not to the extent that E_N lies between E_s and E_d. Varying V_{g4} over a wide range will lead to single-electron current oscillations in the device, with electrons being added one-by-one to the quantum dot.

To operate the device as a turnstile, two r.f. voltages V_{g2} and V_{g3} are applied to gates G2 and G3, of equal amplitude and frequency f, and phase difference π. These voltages alternatively raise and lower the tunnel barriers and modulate their tunnelling probabilities. The cycle starts with the condition in Fig. 5.4(a), but with the tunnel barriers adjusted such that the tunnel probability is low. In the first half of the cycle, the first barrier is lowered and the second is raised, increasing the probability that an electron will tunnel from the source onto the quantum dot (Fig. 5.4[b]). The quantum dot then charges up with a single electron. At the middle of the cycle (Fig. 5.4[c]), the barrier condition is again similar to Fig. 5.4(a), but with an extra electron on the quantum dot. In the second half of the cycle, the second tunnel barrier is lowered and the first is raised, so that the extra electron leaves the quantum dot. The net result in a cycle is the transfer of a single electron per cycle through the quantum dot. The device current is then given by $I = ef$. If the bias V is increased, more than one energy level may lie between E_s and E_d, and the current is given by $I = nef$, where n is the number of levels lying between E_s and E_d. The device I-V characteristics show a current staircase with steps occurring at $I = nef$, shown schematically in Fig. 5.4(e). The slope of the staircase changes as a function of frequency. Kouwenhoven et al. operated their turnstile from 5 to 20 MHz, with an accuracy of ~1%.

The device can also be operated as a single-electron pump, where $I = nef$ even with $V = 0$, or with V opposing the direction of current flow. This operation is possible either by using r.f. gate voltages of unequal amplitudes on G2 and G3, while keeping the phase difference constant at π, or by varying the phase difference but keeping the amplitude constant. Single-electron pumping occurs due to the effect of the gate voltage on the bottom of the potential well of the quantum dot. If this is raised by one of the gate voltages to the extent that an occupied electron level is raised above E_s or E_d, then an electron leaves the quantum dot. With equal gate voltages, this effect is compensated but with unequal voltages, and the effect becomes more likely in one direction, leading to one electron pumped per cycle. With phase differences other than π, the same effect can occur for part of the AC cycle.

5.3 Few-Electron Devices using MTJs

The single-electron turnstiles and pumps discussed in Section 5.2 can transfer one electron per cycle using one or two r.f. voltage signals of frequency f, and current plateaus given by $I = ef$ can be observed in the I-V characteristics. In some devices, e.g. the combined electron turnstile and pump of Kouwenhoven *et al.* (Kouwenhoven *et al.*, 1991b, 1991c), it is possible to transfer controlled numbers of multiple electrons, with $I = nef$. These devices consist of combinations of tunnel junctions and islands, all with significant single-electron charging energies. As a consequence, single-electron charging in the entire device is fundamental to device operation, and leads to an inherent ability to transfer one electron per r.f. cycle through the device.

This section will consider alternative device designs, where MTJs are used to transfer a small number of electrons per cycle of an r.f. signal, in either direction through the device (Tsukagoshi *et al.*, 1997; Jalil *et al.*, 1998; Amakawa *et al.*, 2001). These 'bi-directional electron pumps', originally demonstrated using δ-doped GaAs nanowire MTJs (Tsukagoshi and Nakazato, 1997; Tuskagoshi *et al.*, 1997), have also been investigated in great detail using silicon nanowire MTJs (Altebauemer *et al.*, 2001a, 2001b, 2001c, 2001d, 2002, 2003). Two different designs have been developed, the first using a single r.f. driving signal (Tsukagoshi *et al.*, 1997) and the second using two or more multi-phase r.f. driving signals (Tsukagoshi and Nakazato, 1997; Amakawa *et al.*, 2001).

The electron pump with one r.f. signal uses two MTJs, connected to a central island relatively larger than the islands within the MTJs (Fig. 5.5[a]). The central island is capacitively coupled to a gate electrode and the total capacitance of the central island C_i is usually much larger than the capacitance of the islands in the MTJs. As a consequence, the single-electron charging energy of the central island, given by $E_c \approx e^2/2C_i$, is far smaller than that of the MTJs and may be neglected. The addition of only one electron to the central island is usually not enough to overcome Coulomb blockade in the MTJs. The two MTJs are then effectively decoupled from each other and the central island may be regarded simply as an electron reservoir.

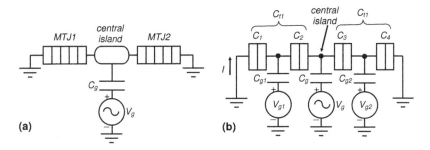

Fig. 5.5 MTJ few-electron pump. (a) Circuit diagram. (b) Simplified circuit, where MTJs are approximated by equivalent double tunnel junctions.

A single r.f. voltage, applied to the gate electrode, is used to drive the circuit. The two MTJs are biased such that they have asymmetric Coulomb gaps and the r.f. voltage overcomes the Coulomb blockade in each MTJ, one after the other, during a cycle. The circuit can then be simply regarded as two clocked MTJs and the current is proportional to the r.f. signal frequency. The number of electrons transferred per cycle depends mainly on the capacitance of the central island and the Coulomb gap of the MTJs, although the RC time of the MTJs becomes important at higher frequency. Both few-electron turnstile and pump operation are possible, and electrons can be pumped in either direction by adjusting the MTJ bias. The operation of these pumps is discussed in detail in Section

5.3.1. The experimental implementation of these pumps in GaAs and in Si is discussed in Sections 5.3.2 and 5.3.3, respectively.

The electron pump with two or more multi-phase r.f. signals uses three or more MTJs in series, with each intermediate island capacitively coupled to an r.f. driving signal. While this device may be simply regarded as an extension of the single r.f. signal pump, its operation is somewhat more complex. The device is of interest for the fabrication of clocked few-electron circuits and for the development of binary decision diagram logic circuits (Chapter 6). The operation and implementation of the pump is discussed in Section 5.3.4.

5.3.1 Operation of single r.f. signal MTJ electron pump

We now discuss the operation of the MTJ electron pump using a charge stability diagram picture, following the theoretical analysis of Jalil *et al.* (Jalil *et al.*, 1998). This approach provides a good understanding of the experimental characteristics of these devices. Using this work, Altebauemer *et al.* (Altebauemer *et al.*, 2001d) have analysed MTJ electron pumps fabricated in silicon in detail (Altebauemer *et al.*, 2001a). We approximate each of the MTJs by simple double tunnel junctions for simplicity (Fig. 5.5[b]). The left and right electrodes of the pump are connected to ground. The single-electron charging levels on the islands of the MTJs can be controlled by means of the gate voltages V_{g1} and V_{g2}, i.e. the biasing point of each MTJ can be adjusted within its stability diagram. The central island is coupled to an r.f. voltage V_g via a gate capacitor C_g. The potential of the central island is φ and the bias (measured from the right-hand side to the left-hand side) across the left and right MTJ is $-\varphi$ and φ, respectively. Assuming that C_g is much greater than the total capacitance C_{t1} and C_{t2} of either arm of the circuit, the potential of the central island φ without any charge build-up is approximately V_g.

The stability diagrams of the MTJs are shown in Fig. 5.6(a). As we have assumed simple double tunnel junctions for the MTJs, the stability diagrams are rhomboid-shaped regions. However, in a more complex MTJ, the stability diagram is also more complex, with large 'main' stability regions separated by smaller stability regions which have complex shapes (Jalil *et al.*, 1998). Even in this case, it is usually possible to find a suitable biasing point for each MTJ. Returning to our simpler, rhomboid-shaped stability regions, we note that the regions are asymmetric about the V_{g1} and V_{g2} axes, due to different tunnel capacitances. This is very likely in actual MTJs. V_{g1} and V_{g2} are adjusted such that each MTJ is biased at different points of their respective stability regions, e.g. MTJ1 is biased nearer the left corner than MTJ2. As a consequence, the Coulomb gaps are asymmetric as a function of the voltage φ across the MTJs.

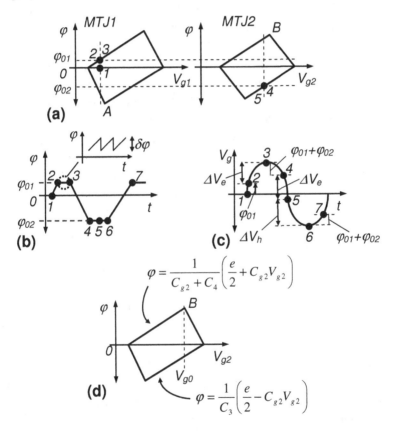

Fig. 5.6 Operation of the MTJ few-electron pump.

The effect of an r.f. voltage V_g with frequency f, applied to the gate electrode, is shown in Fig. 5.6(a–c). At point 1, both MTJ1 and MTJ2 are in Coulomb blockade. As V_g increases, φ initially increases. However, because of the asymmetric biasing of the MTJs, MTJ1 reaches its stability region edge at $+\varphi_{01}$ and comes out of Coulomb blockade first (point 2) while MTJ2 remains in Coulomb blockade. Electrons can now tunnel from the left electrode across MTJ1 onto the central island, as the positive value of φ lowers the central island potential. The central island then charges up and, provided the tunnelling rate is high compared to f, any further increase in V_g does not lead to an increase in φ, with φ pinned approximately at the potential $+\varphi_{01}$. There is a sawtooth oscillation in φ as each electron enters the central node, in a manner similar to the charging of an MTJ trap memory (Nakazato et al., 1994).

As V_g begins to reduce past its positive peak (point 3), the voltage across MTJ1 reduces below $+\varphi_{01}$ and MTJ1 drops into Coulomb blockade. As V_g falls further, MTJ2 reaches the edge of its stability region (point 4) and comes out of Coulomb blockade. This occurs at a point $\varphi_{01} + \varphi_{02}$ below the positive peak of V_g. With further reduction in V_g, the electrons trapped on the central island discharge and all excess electrons are discharged by point 5. Beyond this point, until the negative peak of V_g, electrons leave the central island across MTJ2, leaving 'holes' (i.e. less electrons than in the neutral state) on the central island.

Beyond the negative peak of V_g (point 6), MTJ2 drops again into Coulomb blockade and charge is trapped on the central island. This state is maintained until MTJ1 comes out of Coulomb blockade again at point 7. Note that in Fig. 5.6, we assume that device operation starts from a position where the central island is neutral and the first charging of the central island occurs when $V_g \approx \varphi_{01}$. However, subsequent periods of charging begin when V_g passes a point greater than $\varphi_{01} + \varphi_{02}$ of the maximum or minimum value of V_g.

The net charge pumped in a cycle, i.e. the number of excess electrons and holes on the central island, depends on the widths of the stability regions at the bias points, and on the relatively large capacitance of the central island. For each electron added to the central island, the island potential changes by $\delta\varphi \approx e/C_g$, where we assume that the island capacitance is dominated by C_g. This is the amplitude of the sawtooth oscillation of φ during the charging periods. In Fig. 5.6(b), electrons are added to the central island from point 2 to 3, i.e. for a voltage range ΔV_e.

Therefore, the number of electrons added over this range is $n^+ = \Delta V_e / \delta\varphi$. These electrons discharge from point 4 to point 5, again over a voltage range ΔV_e. Holes are added to the central island from point 5 to 6, i.e. for a voltage range ΔV_p, and the number of excess holes is $p^+ = \Delta V_p / \delta\varphi$. The total number of electrons and holes pumped during the cycle is $N = n^+ + p^+ = (\Delta V_e + \Delta V_h)/\delta\varphi$. Jalil et al. (Jalil et al., 1998) have estimated N for symmetrically chosen values of tunnel junction and gate capacitance, $C_1 = C_4 = 5$ aF, $C_2 = C_3 = 2.5$ aF, $C_{g1} = C_{g2} = 1$ aF and $C_g = 200$ aF. With these values, and with the MTJs biased at the vertices of their stability regions (Fig. 5.6[a], points A and B), $\varphi_{01} = \varphi_{02} = \varphi_0$ and $\Delta V_e = \Delta V_h = \Delta V$. Then $N = 2\Delta V/\delta\varphi \approx 2\Delta V C_g/e = 2(V_0 - \varphi_0)C_g/e$. From the slopes of the sides of the stability region (Fig. 5.6[d]), we can calculate $\varphi_0 = e(C_4+C_{g2})/(C_3(C_4+C_3+C_{g2})) = 0.045$ V. If we apply an r.f. signal with $V_0 = 0.1$ V, the number of electrons and holes transferred per cycle $N = 137$.

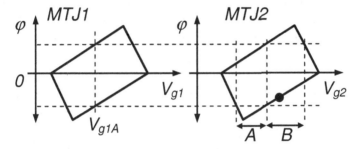

Fig. 5.7 Biasing regions for bi-directional operation of the MTJ few-electron pump.

The MTJ electron pump is bi-directional, and the current direction can be switched by simply readjusting the gate biases V_{g1} and V_{g2}. Figure 5.7 shows this using the MTJ stability diagrams. MTJ1 is biased at a constant gate voltage V_{g1A} and MTJ2 is biased either in region A or in region B. For the MTJ2 biasing point in region A, MTJ2 turns 'on' for positive values of φ, i.e. in the first half of the r.f. cycle, before MTJ1. In contrast, MTJ1 turns 'on' for negative values of φ, i.e. in the second half of the r.f. cycle, before MTJ2. This implies that a net charge is transferred from right to left across the pump. Conversely, for the MTJ2 biasing point in region B, MTJ1 turns 'on' for positive values of φ, i.e. in the first half of the r.f. cycle, before MTJ2, and MTJ2 turns 'on' for negative values of φ, i.e. in the second half of the r.f. cycle, before MTJ1. This implies that a net charge is transferred from left to right across the pump. This method of switching the direction of the pumped current can be contrasted to that in the single-electron pump of Pothier *et al.* (Pothier *et al.*, 1991) or Kouwenhoven *et al.* (Kouwenhoven *et al.*, 1991b), where the phase difference between two r.f. signals is reversed.

Finally, we comment briefly on the temperature and the frequency range of operation of the MTJ electron pump. We consider the device with a requirement that the number of pumped electrons is controlled within one of the expected number N. Considering the temperature first, above a certain maximum temperature T_m, a thermally activated current begins to flow through the MTJs within their Coulomb gap and it is not possible to reliably trap electrons on the central island during an r.f. cycle. However, the trapping of electrons on the central island does not depend on the total width of the Coulomb gap of the MTJs – as for much of the r.f. cycle, the MTJ is biased near the edge of the Coulomb gap, i.e. within $\delta\varphi \approx e/C_g$ of it. The energy barrier preventing unwanted tunnelling only has a height $\sim e\delta\varphi \approx e^2/C_g$ and requiring this to be large in comparison with the thermal fluctuations k_BT, we obtain, for a typical value of $C_g = 200$ aF, the requirement that the operating temperature $T < T_m = e^2/k_BC_g \approx 9$ K. This implies that for 'ideal' pump operation, the operating temperature is rather low. The temperature for ideal operation is clearly limited by C_g and a reduction in this would increase the operating temperature.

Regarding the frequency of operation, at high frequencies, the increased likelihood of missed tunnelling events and at low frequencies, the loss of charge through thermally activated tunnelling, prevents 'ideal' pump operation. We consider the high-frequency limit f_{max} first. By comparing the frequency to the time period Δt of the sawtooth oscillation of φ in Fig. 5.6(d), i.e. the time taken for one electron to charge the central island, Jalil *et al.* (Jalil *et al.*, 1998) determined that $f_{max} = e/4V_0R_tC_g^2$, where R_t is the tunnel resistance. For $R_t \sim 250$ kΩ, i.e. 10 times the quantum of resistance $R_k = h/e^2 = 25.9$ kΩ, and for the capacitance and r.f. signal amplitude used earlier in this section, f_{max} ~40 MHz. Using similar considerations, but comparing Δt to the probability of thermally activated tunnelling, Jalil *et al.* determined that the minimum operating frequency f_{min} is approximately zero if the operating temperature $T \ll T_m$, but rises rapidly as T approaches T_m. This leads to a limit of $T \sim T_m/100$ for 'ideal' pump operation to be possible. For the capacitance values used earlier in this section, $T_m \approx 9$ K and $T \sim 100$ mK.

The MTJ bi-directional electron pump appears not to compare very well in terms of accuracy with true single-electron pumps of the type developed by Pothier *et al.* (Pothier *et al.*, 1991) or Kouwenhoven *et al.* (Kouwenhoven *et al.*, 1991b), and at best, single-electron accuracy may be possible only at milli-Kelvin temperatures. This is a consequence of the large central island size and the large gate capacitance. However, the MTJ pump is simpler to operate, needing only one r.f. signal, and the current direction can be switched by simply changing the DC bias of the MTJs. The device has also been implemented in silicon, and has great potential for the controlled transfer of electrons between different single-electron devices. Furthermore, it is also of great interest as the basic building block of few-electron logic systems, e.g. the 'binary decision diagram' logic scheme of Asahi *et al.* (Asahi *et al.*, 1997).

5.3.2 Single r.f. signal MTJ electron pumps in GaAs

A MTJ electron pump using a single r.f. signal was demonstrated first by Tsukagoshi *et al.* (Tsukagoshi *et al.*, 1997) in δ-doped GaAs material. A schematic diagram of the layout of the device is shown in Fig. 5.8. The device uses two MTJs connected to a central island, with side-gates G1 and G2, and a central gate G coupled to the central island. The circuit diagram of the device is similar to Fig. 5.5(b). The δ-doped layer consisted of a thin plane of Si doping, 30 nm below the material surface. The carrier density and mobility of this layer at 4.2 K were $4 \times 10^{12}/\text{cm}^2$ and $2 \times 10^3/\text{cm}^2/\text{Vs}$, respectively. The MTJs were defined by nanowires, formed by etching the sample below the level of the δ-doped layer. Potential fluctuations created a chain of tunnel barriers and islands along the nanowire, forming an MTJ at 4.2 K (Nakazato *et al.*, 1992). The voltages on the nearby MTJ gates G1 and G2 could be used to modulate the Coulomb gaps of the MTJs. A Coulomb gap with a maximum width

~10 mV was observed in the *I-V* characteristics of each MTJ at 4.2 K. This gap was asymmetric, and could be modulated periodically.

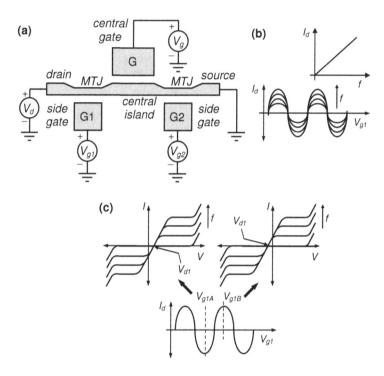

Fig. 5.8 MTJ single-electron pump, using a single r.f. signal V_g. (a) Schematic diagram of device. (b) I_d-V_{g1} characteristics vs. r.f. frequency f. The current I_d increases linearly with frequency f (inset to the upper-right). (c) Drain-source *I-V* characteristics vs. f, at two different values of V_{g1}. The left-hand side curves are biased at V_{g1A}, and the right-hand side curves at V_{g1B}. (after Tsukagoshi *et al.*, 1997).

With an r.f. voltage V_g (−10 dBm, frequency f = 0.3 − 2 MHz) applied to the central gate and a small bias, V_d = 0.1 mV applied to the drain terminal of the device at a temperature of 1.8 K, a drain current I_d could be observed in the device. If an r.f. voltage was not applied, I_d was zero. As a function of the MTJ gate voltages V_{g1} and V_{g2}, oscillations were observed in I_d, corresponding to changes in the biasing point of MTJ1 or MTJ2 within the respective stability region. This is schematically shown in Fig. 5.8(b). Both negative and positive current peaks were observed, as the biasing point of one MTJ (e.g. MTJ1 in the I_d-V_{g1} plot) was shifted and the biasing point of the other MTJ (MTJ2 in the I_d-V_{g1} plot) was kept constant, in a manner similar to the biasing schemes shown in Fig. 5.7. This demonstrated bi-directionality in the operation of the pump. In particular, the negative current peaks corresponded to current flow *against* the effect of the applied drain bias of $V_d = 0.1$ mV, demonstrating true electron pumping. As expected, the negative and positive current peak amplitudes in the oscillations varied linearly with the r.f. frequency, from 0.3 to 2 MHz. Here, the maximum pumped current magnitude was ~50 pA.

Figure 5.8(c) schematically shows the I_d-V_d characteristics of the device, measured at a negative and a positive current peak in the I_d-V_{g1} plot. Current plateaus are observed if an r.f signal is applied, with $I_d \propto f$. The maximum plateau current measured by Tsukagoshi *et al.* (Tsukagoshi *et al.*, 1997) was ~100 pA. If the characteristics are measured at a negative peak, e.g. at V_{g1A}, then zero current is observed in the I_d-V_d characteristics only at a positive value of $V_d = V_{d1}$. This voltage is necessary in order to turn off the pumped current. In a corresponding manner, a negative drain bias $V_d = V_{d2}$ is necessary to turn off the pumped current with the device biased at the negative peak at V_{g1B}. However, it is difficult to observe current plateaus quantized in values of *nef* in the I_d-V_d characteristics at higher than milli-Kelvin temperatures, a consequence of the large capacitance of the central island of the device.

5.3.3 Single r.f. signal MTJ electron pumps in silicon

The single r.f. signal MTJ electron pump developed by Tsukagoshi *et al.* (Tsukagoshi *et al.*, 1997) has been implemented in SOI material by Altebaeumer *et al.* (Altebaeumer *et al.*, 2001a). Altebaeumer *et al.* fabricated the device using MTJs defined by silicon nanowire SETs with side-gates (Smith and Ahmed, 1997), discussed in detail in Chapter 3. Figure 5.9 shows an SEM image of a device. Two side-gated Si nanowire SETs can be seen, defined on either side of a central island, capacitively coupled to an in-plane gate electrode. In their initial devices Altebaeumer *et al.* (Altebaeumer *et al.*, 2001a) used SOI material with a 40 nm-thick top Si layer, heavily-doped to $2 \times 10^{19}/cm^3$ using phosphorous. The BOX was 400 nm-thick. Silicon nanowires ~500 nm in length and ~35 nm in width were defined using e-beam lithography. The wire width was subsequently reduced by dry oxidation to ~25 nm. In subsequent work (Altebaeumer *et al.*, 2001b), the nanowire length was reduced to ~80 nm, to improve the single-electron oscillations in the device.

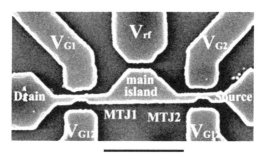

Fig. 5.9 Scanning electron micrograph of an MTJ electron pump, in SOI material. (Reprinted with permission from Altebaeumer and Ahmed, *J. Appl. Phys.* **90**, 1350 [2001]. Copyright 2001, American Institute of Physics).

An MTJ is created along a Si nanowire by potential fluctuations associated with variation in the doping and surface carrier depletion (Chapter 3). In 500 nm-long nanowires (Altebaeumer *et al.*, 2001b), clear, periodic single-electron oscillations were observed in the nanowire current as a function of the side-gate voltage, through there was relatively large variation in the peak and valley currents. In 80 nm-long nanowires, the quality of the oscillations was improved, with more regular peak heights and zero current in the valleys between the peaks (Altebaeumer *et al.*, 2001b). This behaviour corresponded to more regular charge stability regions associated with a less complex MTJ, leading to more regular electron pump characteristics.

Altebaeumer *et al.* observed electron pump operation in the single-electron oscillations of the nanowire current at 4.2 K, with the application of a sinusoidal r.f. signal with frequency f up to 2.5 MHz, and amplitude of 1.2 V. Both positive and negative current peaks were observed, demonstrating bi-directional operation, and it was possible to pump current against an applied DC bias. The height of the current peaks was proportional to the r.f. frequency f. It was also possible to observe current plateaus in the *I-V* characteristics of the device similar to Fig. 5.8(c) (Altebaeumer *et al.*, 2001d), although quantization of the current in multiples of *ef* could not be observed at the measurement temperature of 4.2 K.

The experimental characteristics of the pump could be explained qualitatively using single-electron Monte Carlo simulation (Altebaeumer *et al.*, 2001c). For a typical device, analysis of the experimental characteristics of the SETs allowed an extraction of the various capacitances in the device. For a circuit model similar to that shown in Fig. 5.5(b), Altebaeumer *et al.* extracted $C_1 \approx C_2 \approx 7$ aF, and $C_3 \approx C_4 \approx 5$ aF from the slopes of the stability diagrams of the SETs. A 2-D capacitance model was used to calculate the gate capacitance C_g. Here, $C_g \approx 70$ aF and the total central island capacitance $C_i \approx 150$ aF. These values could then be used for a single-electron Monte Carlo simulation of the pump characteristics, with good qualitative agreement between simulation and experiment (Altebaeumer *et al.*, 2001c). Further works have investigated the behaviour of the pump in a regime where the current varies non-linearly with r.f. signal frequency and amplitude (Altebaeumer *et al.*, 2002), and investigated the effect of cross-capacitances between the gate electrodes and the MTJs on the pump characteristics (Altebaeumer *et al.*, 2003).

5.3.3.1 Device fabrication and experimental characteristics

Figure 5.10 shows the experimental characteristics at 4.2 K from two different Si nanowire pumps, similar to the device of Fig. 5.9 (He *et al.*, 2004b). The pumps use side-gated Si nanowire SETs, defined in the top Si layer of SOI material. The Si layer thickness was 40 nm, and the layer was doped *n*-type (doping concentration $2 \times 10^{19} \text{cm}^{-3}$) with Phosphorus. The buried oxide under this layer was 400 nm thick. The device fabrication process was similar to the Si nanowire SET fabrication process discussed in Chapter 3, but uses a metal etch mask rather than PMMA resist. The first stage of the fabrication process was the definition of mesa regions in the top Si layer, using optical lithography and RIE in 1:1 $SiF_4:SiCl_4$ plasma at 300 W.

After mesa fabrication, Cr/Ag alignment marks were defined on the mesa, for use in the following e-beam lithography stage. Next, a 30 nm-thick Al etch mask was deposited to define the device pattern, using e-beam lithography in PMMA resist, and lift-off of the excess metal. Each SET pattern consisted of a 50 × 100 nm nanowire, with side-gates. The device pattern was then transferred into the silicon by RIE in $SiF_4:SiCl_4$ plasma.

The use of an Al etch mask rather than PMMA allows a deeper etch, deep into the buried oxide layer, and better isolation of the SET gates from the nanowire. Next, the Al mask was removed using Shipley MF319 optical developer to wet etch the Al. Further wet etches were then used to remove the Cr/Ag alignment marks (Ag etch using 1:1 $NH_3:H_2O_2$, followed by a Cr etch using 41g:10.5 ml:250 ml Ceric Ammonium Nitrate:Perchoric Acid:H_2O). Any remaining traces of metal were then removed and the chip degreased using an RCA cleaning etch. The device was then oxidized at 1000°C for 45 minutes, to reduce the nanowire cross-section from 50 to ~20 nm and passivate the etched surfaces. Finally, optical lithography was used to define the peripheral Cr/Ag contacts and bond pads.

Figure 5.10(a–b) shows the pump current I as a function of one of the side-gate voltages V_{g1}, as the frequency and the amplitude of the r.f. signal is varied respectively. In both graphs, a series of positive and negative current peaks is seen, corresponding to current flow in either direction through the pump. In Fig. 5.10(a), as the signal frequency f is varied from 1 to 5 MHz at constant signal amplitude, the amplitude of the current increases. The dependence of I on f, measured at $V_{g1} = -18.6$ V, is shown in the inset to Fig. 5.10(a). This is linear and follows the standard pump equation $I = nef$. Here, the electron packet size n is independent of frequency.

Figure 5.10(b) shows the pump current as the peak–peak signal amplitude V_{p-p} is varied from 0.1 to 0.3 V, at a constant frequency $f = 3$ MHz. As the amplitude of the r.f. signal increases, n increases because the SETs operate at larger source-drain voltages. Following $I = nef$, the current peaks then increase in magnitude linearly. The value of n as a function of V_{p-p}, measured at $V_{g1} = -17.7$ V, is shown in the inset to Fig. 5.10(b). At $V_{p-p} = 0.1$ V, an electron packet only ~10 electrons in size is pumped through the circuit. For very small $V_{p\text{-}p}$, the potential of the centre node falls to an extent that all the SETs are in Coulomb blockade and the current falls to zero.

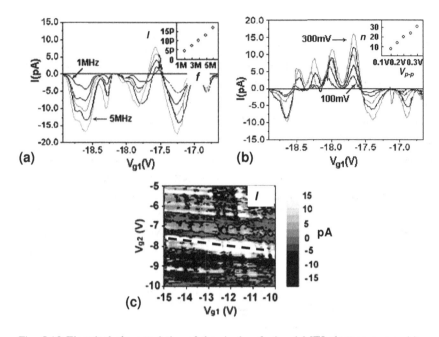

Fig. 5.10 Electrical characteristics of the single r.f. signal MTJ electron pump. (a) Drain current-gate voltage I-V_{g1} characteristics vs. frequency f. (b) I-V_{g1} characteristics vs. peak–peak r.f. voltage $V_{p\text{-}p}$. (c) I vs. V_{g1} and V_{g2} characteristics at constant f. (Reprinted with permission from He et al., Appl. Phys. Lett. **85**, 308 [2004]. Copyright 2004, American Institute of Physics).

Figure 5.10(c) shows the pumped current I in grey-scale at 4.2 K, vs. the gate voltages V_{g1} and V_{g2} of SET 1 and SET 2, respectively. The measurement is from a different device to that of Fig. 5.10(a–b). Here, f = 3 MHz and $V_{p\text{-}p}$ = 200 mV. V_{g1} and V_{g2} are varied from –10 to –15 V and –5 to –10 V, respectively. I oscillates in magnitude if either gate voltage is varied, depending on the position of the operating point of the SETs within their respective stability regions. Both positive (light areas) and negative (dark areas) of current are seen, demonstrating bi-directional electron transfer through the circuit. The graph shows a grid-like pattern, as both V_{g1} and V_{g2} modulate the current in a broadly similar manner. We also note that the current peaks lie along lines (e.g. the dotted line in Fig. 5.10[c]) at a slight angle to the horizontal or vertical. This is due to the capacitive coupling of V_{g1} to SET 2, and V_{g2} to SET 1.

5.3.4 MTJ electron pump with multi-phase r.f. signals

Tsukagoshi and Nakazato (Tsukagoshi and Nakazato, 1997) have developed an electron pump in δ-doped GaAs material using two multi-phase r.f. voltage signals and three MTJs. The circuit diagram of the device is shown in Fig. 5.11(a). Three δ-doped GaAs nanowire MTJs are connected in series, with two intermediate islands coupled by gate capacitors C_{G1} and C_{G2} to two r.f. voltage signals V_{G1} and V_{G2}, respectively. V_{G1} and V_{G2} consist of trapezoidal, triangular or slightly tapered square pulses. The trapezoidal form is shown in Fig. 5.11(b), with pulse duration a third of the pulse train period. The peak amplitude of the pulse V_p is high enough to overcome the MTJ Coulomb blockade.

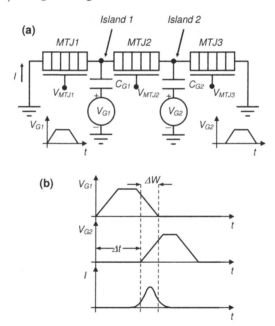

Fig. 5.11 MTJ electron pump with phase-delayed pulses (Tsukagoshi and Nakazato, 1997). (a) Circuit diagram. (b) Operation of the pump, shown schematically.

In the arrangement shown in Fig. 5.11(b), with an overlap between the two pulses, a pumped current is observed in the device current I. This current only takes place if an overlap occurs between the two pulses. If I is measured as a function of the pulse delay Δt, then a current peak is observed for values of Δt corresponding to a pulse overlap. For values of Δt where there is no overlap, at least one MTJ is in Coulomb blockade. For values of Δt where there is an overlap, electrons are transferred across an MTJ if the bias across it is greater than the Coulomb gap. As the two pulses increase and decrease in magnitude, electrons are exchanged at various times between the left electrode and the first island, between the first and second island, and between the second electrode and the right island in a complex manner. However, the net result is electron transfer in one direction.

This design of MTJ pump has been theoretically investigated using stability diagrams by Amakawa *et al.* (Amakawa *et al.*, 2001). Amakawa *et al.* have also discussed the extension of the pump to a chain of MTJs, with multiple r.f. signals. In the following, we explain the operation of the three MTJ pump of Fig. 5.11(a) using the method of Amakawa *et al.* To simplify the analysis, we use triangular pulses V_{G1} and V_{G2} applied to G1 and G2 (Fig. 5.12[a]). We assume that the pump has reached a steady state, where some charge may remain trapped on average on the two islands. This implies that the island potentials φ_1 and φ_2 may have negative values. We also assume that the three MTJs are biased at optimal points of their stability regions. If this is not done then during parts of the cycle, electrons can be transferred in the opposite direction to the desired pumping direction (see Amakawa *et al.*). As V_{G1} and V_{G2} vary, φ_1 and φ_2 vary, and the operating points of the three MTJs shift within the respective stability regions (Fig. 5.12[b]).

Initially, as V_{G1} increases (Fig. 5.12[a]) φ_1 increases from point A towards point B. At point B, MTJ1 reaches the edge of its stability region and turns 'on'. The biasing of MTJ2 is chosen such that for φ_1 corresponding to point A, MTJ2 remains within its stability region and is 'off'. Electrons then transfer across MTJ1 onto Island 1. The potential of Island 1 is then pinned at φ_1 until V_{G1} increases to its peak. When V_{G1} decreases past this point, MTJ1 drops within its stability region and turns 'off'. As MTJ2 is also 'off', electrons remain trapped on Island 1.

However, as V_{G2} also begins to increase, φ_2 increases and the operating point of MTJ2 follows the trajectory B → C. MTJ2 is 'off' until point C is reached. Then MTJ2 turns 'on' and the electrons trapped on Island 1 are transferred to Island 2. At point C, MTJ3 remains 'off' and electrons cannot transfer to the right electrode. However, as V_{G2} reduces, φ_2 also reduces and MTJ3 turns 'on' at point D. Electrons are then transferred from Island 2 to the right electrode. The net result is the pumping of electrons from the left electrode to the right electrode. Figure 5.12(c) shows the movement of electrons through the device as the MTJ operating point moves from A to D.

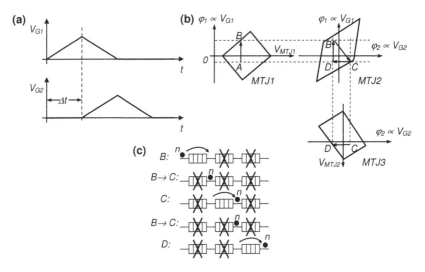

Fig. 5.12 Operation of a three-MTJ electron pump with phase-delayed pulses (after Amakawa *et al.*, 2001).

5.4 Single-Electron Transfer Devices in Silicon

This section discusses two further designs of electron pumps in silicon, where it is possible to transfer one, or an integer number of electrons, using two AC signals. These devices are based on nanoscale CCDs (Fujiwara *et al.*, 2001, 2004), and on SETs combined with nanoscale MOSFETs (Ono and Takahashi, 2003; Ono *et al.*, 2003).

5.4.1 Single-electron transfer using a CCD

Fujiwara et al. (Fujiwara et al., 2004) have fabricated a CCD using a silicon nanowire, where it is possible to transfer single electrons in each cycle of a series of two phase-shifted voltage pulses, forming a single-electron turnstile. The device was based on earlier work by Fujiwara et al. (Fujiwara et al., 2001) on Si nanowire CCDs, where it was possible to manipulate single holes, generated using halogen lamp or He-Ne laser illumination, between two potential wells. A schematic diagram of the device is shown in Fig. 5.13(a). The device consists of a 30 nm-wide nanowire defined in the top Si layer of SOI material. The thickness of the Si layer is 20 nm. Two polysilicon gates G1 and G2 are defined above the nanowire, supported on a gate oxide 30 nm thick. The gate length for both G1 and G2 is 40 nm. G1 and G2 can be used to control the current in the underlying section of nanowire and define two nanoscale MOSFETs. A further deposited upper gate, which fills the region between G1 and G2, can be used to control the potential of the section of the nanowire between G1 and G2. With a low voltage V_{off} applied to G1 and G2, the underlying nanowire channel is depleted and MOSFET1 and MOSFET2 are 'off'. A higher voltage V_{on} can be used to turn the MOSFETs 'on'.

The device operates using a series of pulses V_{g1} and V_{g2}, applied to G1 and G2 as follows (Fig. 5.13[b]). A small potential V_d is applied to the drain of the device. For the first part of the pulse cycle $t_1 \rightarrow t_2$, MOSFET1 and MOSFET2 are 'off' and the intermediate section of nanowire forms an island isolated by wide potential barriers. The island size is small enough such that it has an estimated total capacitance of only ~10 aF and at low temperature, discrete single-electron charging levels exist on the island. From $t_2 \rightarrow t_3$, when V_{g1} changes from V_{off} to V_{on}, MOSFET1 turns 'on' and electrons can flow between the source and the island. At t_3, MOSFET1 turns 'off' and a small number of extra electrons n_1 (one extra electron in Fig. 5.13[b]) are trapped on the island. Here, MOSFET2 remains 'off', isolating the island from the drain and, as a consequence, V_d has little effect on the island potential. This is very different from the single-electron turnstile discussed in Section 5.2.1. The number of electrons trapped on the island is given by $n_1 - \frac{1}{2} < C_g V_{ug}$

$< n_1 + \frac{1}{2}$, where V_{ug} is the upper gate voltage and C_g is the upper gate to island capacitance. From $t_4 \rightarrow t_5$, MOSFET2 turns 'on' and a small number of electrons n_2 leave the island. This number is given by $n_2 - \frac{1}{2} < C_g(V_{ug} - V_d) < n_2 + \frac{1}{2}$ and is different from n_1 due to the effect of the drain voltage. From $t_5 \rightarrow t_6$, both MOSFETS are 'off' again, completing the pulse cycle. The net result is the transfer of $n_1 - n_2$ electrons per cycle through the device, with a current given by $I = (n_1 - n_2)ef$.

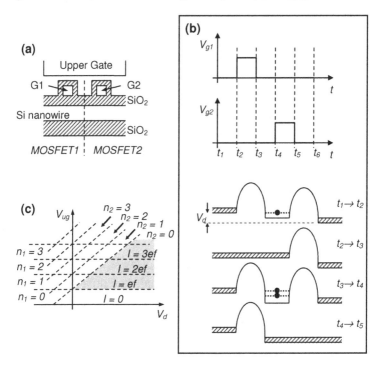

Fig. 5.13 Electron pump using a CCD (after Fujiwara *et al.*, 2004). (a) Schematic diagram. (b) Electrical operation. (c) Stability diagram.

Fujiwara *et al.* (Fujiwara *et al.*, 2004) analyse the operation of this device using a stability diagram picture (Fig. 5.13[c]). The stability diagram may be drawn as a function of V_{ug} and V_d, and shows the numbers of electrons n_1 and n_2 entering and leaving the island, respectively. As discussed above, n_1 only depends on V_{ug} and does not depend on V_d. The regions $n_1 = 1, 2, 3...$ are then parallel to V_d. In contrast, n_2 depends on both V_{ug} and V_d and the regions $n_2 = 1, 2, 3...$ lie diagonally across the diagram. A pumped current is seen in the regions where $n_1 - n_2 > 0$ (shaded area). The current through the device is given by $I = (n_1 - n_2)ef$, and as a function of V_{ug}, forms a staircase as $n_1 - n_2$ increases in integer values.

High-frequency operation is possible as electrons enter and leave the island with the potential barriers lowered and Fujiwara *et al.* (Fujiwara *et al.*, 2004) have operated the device up to a frequency of 100 MHz. The device is also capable of operating at a relatively higher temperature of 20 K. The error rate was expected to be less than 10^{-2} at 100 MHz. The control of the number of electrons pumped per cycle can be adjusted simply by adjusting V_{ug}. This is very different from the single-electron turnstiles of Geerligs *et al.* (Geerligs *et al.*, 1991), where the number of pumped electrons depends on V_d, and from the single-electron pump of Pothier *et al.* (Pothier *et al.*, 1991), where careful control over the values of V_{g1} and V_{g2} is necessary (see Section 5.2).

While there are similarities with the single-electron turnstile of Kouwenhoven *et al.* (Kouwenhoven *et al.*, 1991c) (Section 5.2.3) in that both devices use the modulation of potential barriers, there are significant differences in that the single-electron turnstile of Kouwenhoven *et al.* uses tunnel barriers rather than wide potential barriers, and only the tunnelling probability through the barriers is modulated.

5.4.2 SET/MOSFET single-electron pump and turnstile

Ono and Takahashi (Ono and Takahashi, 2003, Ono *et al.*, 2003) have developed a single-electron pump and turnstile using a SET combined with two nanoscale MOSFETs. A circuit diagram of the device is shown in Fig. 5.14(a). The device consists of a SET with source and drain regions formed by MOSFETs 1 and 2. Gate voltage V_{g1} and V_{g2} are used to control the channel current through the two MOSFETs. The gates also couple via C_s and C_d to the island of the SET. This device was defined in SOI material, with the SET defined by a PADOX technique (Ono *et al.*, 2000b). This technique is discussed in detail in Chapter 3. The SET total island capacitance was ~5.4 aF and the device was operated at a relatively high temperature of 25 K.

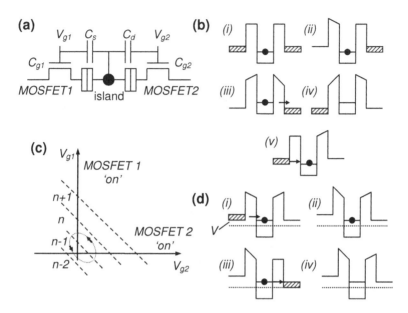

Fig. 5.14 SET/MOSFET single-electron pump and turnstile (after Ono and Takahashi, 2003; and Ono et al., 2003). (a) Circuit diagram. (b) Pump operation. (c) Stability diagram. (d) Turnstile operation.

The device operates as a single-electron pump using AC voltages V_{g1} and V_{g2} as follows. Fig. 5.14(b) shows the potential across the device as V_{g1} and V_{g2} are varied. Here, (i) V_{g1} and V_{g2} are adjusted such that both MOSFETs are 'on' and the SET has a fixed number of electrons n on the island. (ii) The SET source is then depleted of electrons by applying a negative value of V_{g1}, turning MOSFET1 'off'. As V_{g1} also affects the island potential, a compensating positive value of V_{g2} is applied simultaneously. This keeps the island potential approximately constant. (iii) The island potential is then raised so that an electron is transferred to the right electrode. (iv) MOSFET1 is turned 'on' and MOSFET2 is turned 'off', with the island potential constant. (v) The island potential is then lowered by adjusting V_{g1} and V_{g2} such that an electron transfers on to the island from the left electrode. V_{g1} and V_{g2} are then returned to the condition of (i), completing the cycle. The net result is the transfer of one electron from the left to the right electrode.

Ono and Takahashi (Ono and Takahashi, 2003) have analysed the electron pump operation using a stability diagram picture (Fig. 5.14[c]). As a function of V_{g1} and V_{g2}, the number of electrons n is fixed in diagonal regions across the plot, with the slope of these regions determined by the ratio of the coupling capacitances C_s and C_d of the island to the two gates. In addition, for negative or positive values of V_{g1}, MOSFET1 is 'off' or 'on', respectively. MOSFET2 behaves in a similar manner with V_{g2}. If V_{g1} and V_{g2} are varied following the trajectory shown by the dotted circle, the island moves from the state $n \rightarrow n-1 \rightarrow n$, and one electron is transferred across the pump. Reversing the sense of the trajectory allows electrons to be pumped in the opposite direction. Ono and Takahashi have operated this device at 25 K using a frequency of 0.5 MHz and 1 MHz. It was possible to transfer one or two electrons per cycle by controlling V_{g1} and V_{g2} such that either one or two boundaries between stability regions were crossed (Fig. 5.14[c], dashed lines). The corresponding *I-V* characteristics show current plateaus at $\pm ef$ or at $\pm 2ef$.

The device has also been operated as a single-electron turnstile at 25 K, using 0.5 and 1 MHz AC gate voltages (Ono *et al.*, 2003). This operation uses a constant bias V applied between the left and right electrodes. The AC voltages V_{g1} and V_{g2} are phase shifted by π and the device is operated in a regime where at least one of the MOSFETs is 'off' during the cycle. Fig. 5.14(d) shows the potential across the device during the cycle.

Here, (i) MOSFET1 is 'on' and an electron is transferred from the left electrode onto the island. (ii) both MOSFETS are turned 'off' and an electron is trapped on the island. MOSFET2 is then turned 'on'. (iii) the electron moves from the island onto the right electrode. Finally, (iv) both MOSFETs are turned 'off' again. The net result is again the transfer of one electron through the device per cycle. Note that it is not possible to observe a current without AC gate voltages in this regime, as at least one of the MOSFETs is always 'off'.

5.5 Metrological Applications

We briefly consider metrological applications of single-electron turnstiles and pumps. The ability of these devices to transfer only one, or an integer number of electrons, per cycle of an applied AC signal raises the possibility of a current standard, where the current I is related to a signal of known frequency f by the relation $I=ef$ (Odintsov, 1991). A standard for capacitance may also be defined, based on the measurement of the voltage produced by a known charge (Williams *et al.*, 1992). This can be achieved by using a single-electron pump as an electron counter to place an exact number of electrons onto a 1 pF capacitor (Keller *et al.*, 1996).

We now discuss the use of a single-electron pump as an electron counter for metrological device application, e.g. a capacitance standard. Such a standard requires that the error in the electron counting is less than 1 in 10^8, i.e. 10 parts per billion (ppb) (Williams *et al.*, 1992). This requires extremely accurate single-electron pumps, where the chances of an electron not being transferred in a cycle, or more than one electron being transferred in a cycle, is less than 1 in 10^8. Typically, this requires single-electron pumps using MTJs and islands, with each island capacitively coupled to a different r.f. signal of specific phase. A 5-junction single-electron pump with an accuracy of 0.5 parts in 10^6 was developed by Martinis *et al.* (Martinis *et al.*, 1994). An improved device using a 7-junction single-electron pump increased the accuracy to 15 parts in 10^9, the level required for metrological application (Keller *et al.*, 1996).

A circuit diagram of the 7-junction single-electron pump of Keller *et al.* (Keller *et al.*, 1996) is shown in Fig. 5.15. The device uses seven Al/AlO$_x$ tunnel junctions, with six intermediate islands capacitively coupled to gates G1–G6. The pump is connected to an external island with a capacitance C_s ~20 pF. A double tunnel junction SET, used as an electrometer, senses the potential V_i of the external island. A cryogenic temperature switch S is used to measure the current voltage curve of the MTJs, determine an adjustment to the gate voltages to remove cross-capacitances and offset charge effects on the islands, and calibrate the electrometer. During pump operation, S is kept open. Phase-shifted AC signals are then applied to the gates G1–G6, transferring one electron successively through the tunnel junctions onto the external island and then off the external island.

The change in island potential V_i is detected by the electrometer. The inset to Fig. 5.15 schematically shows the change in V_i as a function of time. This forms a square wave pattern, where each switch in V_i corresponds to one electron entering or leaving the external island. Keller *et al.* operated their pump at a frequency of 5.05 MHz at a temperature from 35 to 200 mK. As the operating frequency was faster than the response time of the electrometer, during normal pump operation a constant value of V_i was detected, unless an error, i.e. a missed pumping event or leakage of the charge stored on the external island occurred. Errors created sudden jumps in V_i as the electron number changed. The error count corresponded to an accuracy of 15 ppb, with an average hold time of electrons on the island of 600 s.

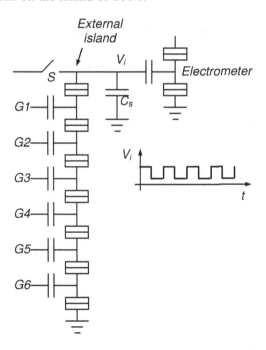

Fig. 5.15 Single-electron pump for metrological applications (after Keller *et al.*, 1996).

The accuracy of single-electron turnstiles and pumps has been theoretically investigated in a number of works. In early work on the accuracy of single-electron pumps, Jensen and Martinis (Jensen and Martinis, 1992) investigated errors caused by thermal activation, co-tunnelling (i.e. the transfer of charge within the Coulomb gap through virtual states on the island), and operating the device at too high a frequency. They calculated the pump parameters needed to obtain metrological accuracies, and concluded that a 5-junction or better pump was necessary.

Averin *et al.* (Averin *et al.*, 1993) also investigated co-tunnelling processes, and calculated an accuracy of 1 part in 10^8 in a 5-junction pump, or a 5-junction per arm (i.e. 10-junction) turnstile. Other error mechanisms, e.g. photon-assisted co-tunnelling, have also been proposed (Martinis and Nahum, 1993). The accuracy of single-electron pumps has been compared to single-electron turnstiles by Fonseca *et al.* (Fonseca *et al.*, 1996a). They found that pumps appear to be substantially better, e.g. even an 8-junction turnstile is not equivalent to a 5-junction pump. Fonseca *et al.* (Fonseca *et al.*, 1996a, 1996b) have also proposed an optimized step-like waveform rather than a triangular waveform to drive turnstiles and pumps. With such a waveform, in a 5-junction electron pump with junction capacitances ~0.1 fF, temperature ~100 mK and frequency 10 MHz, the theoretical error can be as low as 1 in 10^{13}.

Chapter 6

Single-Electron Logic Circuits

6.1 Introduction

Single-electron devices may be used to fabricate circuits where one, or at most a few, electrons are manipulated to perform logic operations. This possibility was identified even in early works on single-electron devices (Averin and Likharev, 1986; Likharev, 1987; Likharev, 1988; Averin and Likharev, 1992). 'Single-electron logic' circuits may be broadly realized in two different ways. One approach is to use SETs as switching transistors, in a manner similar to conventional MOS transistors, for the fabrication of resistively-loaded or complementary logic gates. Such an approach, where high and low voltage levels define the '1' and '0' bits, has been referred to as 'voltage state' logic (Likharev, 1999). An elegant alternative, more suited to the inherent advantages of single-electron devices, is to use a single electron, or at most a few-electron packet, to define a bit. The presence of an electron at a given point in the logic circuit then represents a '1', and the absence of the electron represents a '0'. This approach, where the presence or absence of a small charge packet defines the bits, has been referred to as 'charge state' logic (Likharev, 1999).

A large body of theoretical work exists on the implementation of single-electron logic circuits, using both the voltage state and the charge states approach (See Likharev, 1999). However, through most of the 1990s, there were relatively few experimental implementations of these circuits, especially in comparison with work on single-electron memories (Chapter 4). This was mainly a consequence of the difficulty of fabricating circuits using multiple single-electron devices, all with similar electrical characteristics. Furthermore, the circuits may require single-electron devices with a high level of performance, e.g. most voltage state approaches require gain in the constituent SETs, a difficult requirement to attain. More recently, with advances in nanofabrication techniques leading to control over the device dimensions at the ~10 nm scale, further experimental implementations have been reported (See Ono et al., 2005). These include circuit operation at room temperature.

This chapter will discuss the design and fabrication of single-electron logic circuits in detail, first using the voltage state logic approach, and then using the charge state logic approach. We will emphasize, where possible, circuits in silicon.

6.2 Voltage State Logic

Single-electron transistors can be used as voltage switches to implement basic logic gates. (Averin and Likharev, 1986; Likharev, 1987, 1988; Averin and Likharev, 1992). This approach uses an input voltage, applied to the SET gate terminal, to control the SET source-drain conductance and switch the output voltage across the SET between high and low voltage levels. A high voltage state then defines the '1' bit, and a low voltage state defines the '0' bit. Here, the SET is simply utilized as a circuit element with a particular electrical characteristic, and the ability of the device to control exact numbers of electrons is not used directly. The scheme, usually referred to as 'voltage state' logic (Likharev, 1999), is very similar to conventional MOS-based logic, and analogues of n-MOS, p-MOS or CMOS logic gates are possible. The much smaller size of the SET in comparison to a conventional, 'classical' MOSFET implies that the area occupied by the gate can be considerably reduced. Furthermore, the far smaller electron packet sizes possible with SETs may lead to greatly reduced switching power dissipation.

While both SET and CMOS logic gates use voltage levels to define the bits, and use switching transistors to determine these levels, there are fundamental differences in the characteristics of the two devices. This leads to differences in the detailed design of the logic gates. The primary difference is the oscillatory nature of the SET I_{ds}-V_{gs} (i.e. transfer) characteristic. This is very different from the I_{ds}-V_{gs} characteristic in a MOSFET, where I_{ds} increases monotonically with V_{gs}. An oscillatory I_{ds}-V_{gs} characteristic implies that, depending on the range of gate voltage, I_{ds} may increase or decrease with increasing V_{gs}. From the perspective of the electrical characteristics, the former situation corresponds to an n-MOSFET-like device and the latter to a p-MOSFET-like device. It is then possible to implement n- or p-MOSFET-like devices using the same basic SET design, simply by biasing the SET at different points of the I_{ds}-V_{gs} characteristic, and without using different types of doping.

However, the more complex nature of the SET I_{ds}-V_{gs} characteristic implies greater design complexity. More significantly, the SET has poor voltage gain, at best somewhat greater than one and often less than one, unlike the high gain in a MOSFET. This greatly restricts the ability of a SET logic gate to drive further devices, i.e. 'fan-out' of the gate. The SET also suffers from restrictions on the possible range of input and output voltages. Furthermore, the high sensitivity of the SET to thermal fluctuations implies that the input/output voltage swing reduces rapidly with temperature.

A variety of logic gates are possible using the above approach. These include gates with resistively-loaded SETs, complementary SETs, 'programmable' logic gates, and the use of more than one gate terminal on the same SET to implement logic functions. Each of these approaches is discussed in the following.

6.2.1 SET inverter with resistive load

Perhaps the simplest voltage state logic gate is the SET inverter with a resistive load (Likharev, 1988; Korotkov *et al.*, 1995). Figure 6.1(a) shows a circuit diagram of the device. A double tunnel junction SET is connected to the supply voltage V_{dd} by a resistive load R. An input voltage $V_{in} = V_g$ is capacitively coupled via a gate capacitor C_g to the SET island. We assume that there are zero offset charges and for zero input gate voltage, the SET is biased within the Coulomb blockade. In an experimental realization of this circuit, an additional, capacitively coupled voltage terminal may be necessary to adjust the biasing point of the SET and remove the effect of any offset charges. The output voltage

V_o is measured across the SET and across the output capacitance C_o. Here, C_o can represent the interconnect capacitance, or the input capacitance of the next stage.

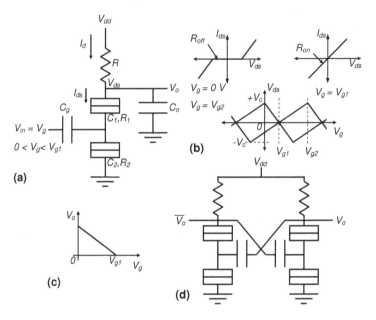

Fig. 6.1 SET inverter with resistive load. (a) Circuit diagram. (b) Stability diagram and schematic I-V characteristics. (c) Transfer characteristics. (d) Latch circuit.

Figure 6.1(b) schematically shows the charge stability diagram of the SET, as a function of the SET gate voltage V_g and the SET drain-source voltage V_{ds}. The diagram shows trapezoidal- or 'diamond'-shaped regions of constant island charge (Chapter 2), and only the regions along the V_g axis are shown for simplicity. Within the regions shown, the SET is in Coulomb blockade and 'off'. Here, the SET drain-source resistance R_{off} is very high at low temperature, and the current I_{ds} is zero (left inset, Fig. 6.1[b]). If V_{ds} and V_g are such that the bias point lies outside these regions, then the SET conducts and is 'on', with a relatively low drain-source resistance R_{on} (right inset, Fig. 6.1[b]). The maximum width of the Coulomb blockade is $\pm V_c$.

For a low input voltage, $V_{in} = V_g = 0$ V, assuming zero initial charge on C_o, $V_o = V_{ds} < V_c$ and the SET is 'off'. Charge then flows from V_{dd} into

C_o, and V_o increases until it reaches the edge of the Coulomb blockade region. In Fig. 6.1(b), this occurs at $V_o = V_c$. C_o cannot charge any further as the SET would then turn 'on', discharging this excess charge to ground. This pins V_o at the edge of the Coulomb blockade region at V_c and the output is 'high'. We note that the supply voltage V_{dd} can be relatively small and does not need to be much greater than V_c. Now, if V_{in} is increased from 0 V to V_{g1}, then V_o follows the edge of the Coulomb blockade region, reducing from V_c to 0 V. Therefore, for 'high' input, i.e. $V_{in} = V_{g1}$, the output is 'low', i.e. $V_o = 0$ V. The circuit then operates as an inverter or 'NOT' logic gate, with the transfer characteristic shown in Fig. 6.1(c).

When the output is 'high' and the SET is 'off', a real SET operating at a non-zero temperature has a high but finite resistance R_{off}. This implies that $V_o = R_{off}V_{dd}/(R + R_{off})$ and to obtain $V_o \sim V_c$, it is necessary for $R_{off} >> R$. Similarly, when the output is 'low', $V_o = R_{on}V_{dd}/(R + R_{on})$ and to obtain $V_o \sim 0$ V, it is necessary for $R_{on} << R$. This limits R to the range $R_{on} << R << R_{off}$.

It is straightforward to extend this design to a flip-flop or latch (Fig. 6.1[d]), with two inverters connected back-to-back (Korotkov et al., 1995). However, the device does require that each inverter has a voltage gain $\Delta V_o/\Delta V_g > 1$. This is possible for the operating range of the SET corresponding to the section of the stability diagram from $V_g = 0$ V to V_{g1}, where the edge of the Coulomb blockade along V_{ds} reduces with increasing V_g. For this region, the gain is inverting and given by $\Delta V_{ds}/\Delta V_g = -C_g/C_2$ (Likharev, 1987). Increasing the gate capacitance helps in obtaining a voltage gain greater than one. We note that a variety of semiconductor and metal-based SETs have been fabricated with greater than unity gain (Zimmerli et al., 1992; Visscher et al., 1994; Smith and Ahmed, 1997b; Ono et al., 2000c; Heij et al., 2001).

6.2.2 Complementary SET inverter

We now consider the operation of the complementary SET (C-SET) inverter (Tucker, 1992). Figure 6.2(a) shows the circuit diagram of the device. The device consists of two SETs in series, coupled by gate capacitors C_g to the input voltage $V_{in} = V_g$. The output voltage V_o is measured at the point between the SETs, across output capacitance C_o. The upper SET (SET1) is biased using a trimming gate V_{t1} such that it operates in a manner analogous to a p-MOSFET. Conversely, the lower SET (SET2) is biased using a trimming gate V_{t2} such that it operates in a manner analogous to an n-MOSFET.

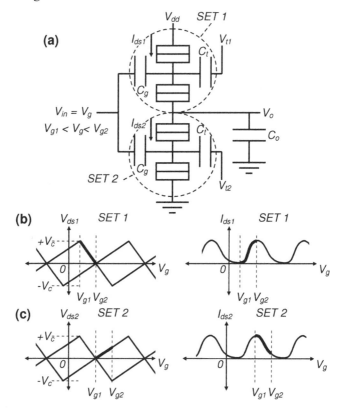

Fig. 6.2 Complementary SET inverter. (a) Circuit diagram. (b–c) Stability diagram and biasing with respect to the single-electron oscillation characteristics of SET1 and SET2, shown schematically.

The operation of the device is first considered using a simple picture where the input voltage turns the SETs 'on' or off'. Figures 6.2(b–c) schematically show the charge stability diagram and single-electron oscillations for SET1 and SET2, respectively. SET1 is biased along a section of the stability diagram from V_{g1} to V_{g2}, where the Coulomb blockade region increases with V_{in}. As V_{in} increases from V_{g1} to V_{g2}, the

SET drain-source current I_{ds1} reduces and the SET begins to turn 'off'. SET1 then operates in a manner analogous to a p-MOSFET. Conversely, SET2 is biased along a section of the stability diagram, from V_{g1} to V_{g2}, where the Coulomb blockade region reduces with V_{in}. Therefore, as V_{in} increases from V_{g1} to V_{g2}, the SET drain-source current I_{ds1} increases and the SET turns 'on'. SET1 then operates in a manner analogous to an n-MOSFET.

For 'low' input, $V_{in} = V_{g1}$, SET1 may be replaced by the corresponding 'on' resistance R_{on}, and SET2 by the corresponding 'off' resistance R_{off}. The output voltage is then given by $V_o = R_{off}V_{dd}/(R_{off} + R_{on}) \sim V_{dd}$, i.e. the output is pulled 'high'. Conversely, for 'high' input, $V_{in} = V_{g2}$, SET1 may be replaced by the corresponding 'off' resistance R_{off}, and SET2 by the corresponding 'on' resistance R_{on}. Therefore, the output voltage is given by $V_o = R_{on}V_{dd}/(R_{off} + R_{on}) \sim 0$ V, i.e. the output is pulled 'low'. The circuit then operates as a complementary inverter.

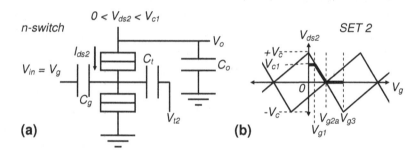

Fig. 6.3 Complementary SET inverter: The n-switch. (a) Circuit diagram. (b) Stability diagram.

We now discuss the switching of the C-SET inverter device in more detail, where we include the charging of the output capacitance C_o. Following the analysis of Tucker (Tucker, 1992), we break the full inverter circuit into two parts, an 'n-switch' formed by SET2 and C_o, and a 'p-switch' formed by SET1 and C_o. We first consider the operation of the n-switch. Figure 6.3(a) shows the circuit diagram of the n-switch and Fig. 6.3(b) shows the charge stability diagram of SET2. Here, the maximum value of the Coulomb blockade is $\pm V_c$. The 'trimming' gate V_{t1} is used to move the position of the trapezoidal charge stability regions along the V_g axis, to obtain the situation illustrated in Fig. 6.3(b). Here, the region from V_{g1} to V_{g2} corresponds to the negative slope part of the first trapezoidal region. We assume that for 'low' input, $V_{in} = V_g = 0$ V, V_o is 'high' and is charged to a value $V_{c1} < V_c$. The SET is then 'off' and within the first trapezoidal region.

Now, if V_{in} is increased, the bias point follows the solid line in Fig. 6.3(b), and the SET remains 'off' until $V_{in} = V_{g1}$. If V_{in} is increased beyond V_{g1}, the SET turns 'on', discharging C_o and reducing V_o until it reaches the edge of the Coulomb blockade region. In a manner similar to the resistively-loaded inverter, V_o is then pinned at the edge of the Coulomb blockade region. As V_{in} is increased further, C_o discharges until at $V_{in} = V_{g2a}$, $V_o = 0$ V and C_o is fully discharged. If V_{in} is increased further to V_{g3}, $V_{ds2} = V_o = 0$ V implies that the SET turns 'off' because the bias point moves within the next trapezoidal region.

V_o then follows the path along the solid line in Fig. 6.3(b), into this region. Therefore, for 'high' input, $V_{in} = V_{g3}$, the output is 'low', i.e. $V_o = 0$ V and the circuit operates as an inverter. We note that a quasi-static switching process may be necessary to allow C_o to discharge fully, otherwise the SET bias point can move into the second trapezoidal stability region, turning the SET 'off' and preventing a complete discharge of C_o.

We next consider the operation of the p-switch, formed by SET1 and C_o. Figure 6.4(a) shows the circuit diagram of the switch, Fig. 6.4(b) shows the charge stability diagram of SET1 and Fig. 6.4(c) shows the transfer characteristic of the switch. Here, SET1 is connected to the supply voltage $V_{dd} = V_{c1}$. The 'trimming' gate V_{t2} is used to move the position of the trapezoidal charge stability regions along the V_g axis and create the situation illustrated in Fig. 6.4(b), such that the region from V_{g2b} to V_{g3} corresponds to the positive slope part of the first trapezoidal region. We assume that for 'high' input, $V_{in} = V_g = V_{g3}$, the output voltage is 'low', $V_o = 0$ V. This implies that the SET drain-source voltage $V_{ds1} = V_{dd} - V_o = V_{c1}$. The SET is then 'off' and within the second trapezoidal charge stability region, and C_o is disconnected from V_{dd}.

Single-Electron Logic Circuits 217

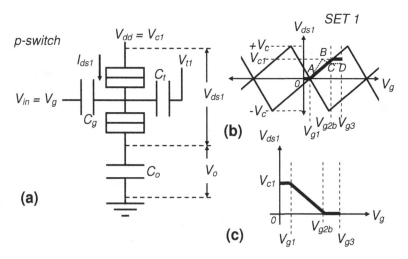

Fig. 6.4 Complementary SET inverter: The p-switch. (a) Circuit diagram. (b) Stability diagram. (c) Transfer characteristics.

Now, if V_{in} is reduced, the bias point follows the solid line in Fig. 6.4(b), the SET remaining 'off' until $V_{in} = V_{g2b}$. At this point, the SET turns 'on' and C_o can charge up, increasing V_o. This process reduces $V_{ds1} = V_{dd} - V_o$ to less than V_{c1} until the edge of the stability region is reached and the SET goes into Coulomb blockade and turns 'off'. As V_{in} is reduced to V_{g1}, V_{ds1} is pinned at the edge of the Coulomb blockade region, and reduces (solid line, Fig. 6.4[b]) as C_o charges up. At $V_{in} = V_{g1}$, $V_{ds1} = 0$ V and C_o is fully charged with $V_o = V_{c1}$. As V_{in} is reduced to 0 V, the SET enters the first charge stability regions and turns 'off'. Therefore, for 'low' input, $V_{in} = 0$ V, the output is 'high', i.e. $V_o = V_{c1}$ (Fig. 6.4[c]) and the circuit operates as an inverter. Again, a quasi-static switching process may be necessary to correctly operate the switch.

Connecting the n-switch and p-switch together forms the full C-SET inverter (Fig. 6.2). The transfer characteristic of the full inverter is shown in Fig. 6.5, obtained by the overlap of the individual transfer characteristics of the two switches. Here, we have assumed for simplicity that $V_{g2} = V_{g2a} = V_{g2b}$, i.e. the charge stability regions of the SETs have slopes of similar magnitude along all sides. The two SETs then turn 'on' and 'off' at similar points, and in the shaded regions, both SETs are 'off'.

While this tends to isolate the output from both ground and the supply terminal, the output is driven 'high' or 'low' in a quasi-static switching process. For a more general case, $V_{g2a} \neq V_{g2b}$ (illustrated in Figs 6.3 and 6.4) it may not be possible to discharge C_o fully, reducing the voltage swing of V_o to less than V_{c1}. To demonstrate this, we consider V_{in} increasing from 0 V to V_{g3}.

At V_{g1}, both SET1 and SET2 tend to turn 'on'. Assuming SET2 has a lower 'on' resistance than SET1, above V_{g1} SET2 discharges C_o faster than SET1 can charge up C_o, driving SET1 along the line AB in Fig. 6.4(b). At V_{g2a}, SET2 turns 'off' (Fig. 6.3[b]) and above this voltage, SET1 can charge up C_o as long as SET2 does not turn 'on'. The operating point of SET1 then follows the line BC, with the SET turning 'off' at C. For a further increase in V_{in} to V_{g3}, both SET1 and SET2 are 'off'. However, the maximum value of V_{ds1} is reduced to less than V_{c1}, implying that the output voltage $V_o = V_{c1} - V_{ds1}$ is not zero for high input voltage. In Fig. 6.5, the transfer characteristic follows the line BCD.

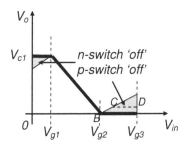

Fig. 6.5 Complementary SET inverter transfer characteristic.

Korotkov *et al.* (Korotkov *et al.*, 1995) have theoretically analysed the effect of temperature and variation in offset charge on the performance of both the resistively-loaded and C-SET inverter. This work suggests that the performance of these devices is strongly dependent on the value of the offset charge. However, even for an optimum value of this charge, the maximum operating temperature T_{max} is relatively low, $\sim 0.025 e^2/C k_B$, where C is the total island capacitance. For $C \sim 0.5$ aF, T_{max} is only ~ 100 K. Room temperature operation would require $C \sim 0.1$ aF, corresponding to islands only a few nanometers in size. For wider parameter margins, the operating temperature drops and for realistic circuits, T_{max} is only ~$0.01e^2/Ck_B$.

A C-SET inverter operating at 27 K has been fabricated in SOI material by Ono et al. (Ono et al., 2000c). The device used two SETs defined by vertical pattern-dependent oxidation (V-PADOX) (Ono et al., 2000a), where oxidation of a thinned region of the top silicon layer of the SOI material led to the formation of two single island SETs (See Chapter 3 for details of this method). The process created islands of similar size and high-resolution lithography was not necessary to directly define the islands. Two SETs could be packed within an area of only 200 nm × 100 nm. Each SET used a side-gate to adjust the operating point, such that one of the SETs behaved as an *n*-switch and the other as a *p*-switch. A top-gate was coupled to both SETs, with a capacitance of C_g ~2 aF to each SET. The top-gate was used as the input terminal to the inverter. As the SET source and drain tunnel capacitances, C_1 and C_2, respectively, were ~1 aF, the maximum voltage gain of each SET was C_g/C_2 ~ 2. Ono et al. measured the transfer characteristic of the C-SET at 27 K and found that this was a compromise between a high output voltage swing and a voltage gain, with the best value of voltage gain ~1.3. Increasing the supply voltage increased the voltage gain such that it was closer to the gain in the SETs. However, a higher supply voltage led to a higher co-tunnelling current, increasing the valley current in the single-electron oscillations and reducing the voltage swing.

6.2.3 Complementary SET NAND and NOR gates

The concepts used in the C-SET inverter can be extended to implement C-SET NAND and NOR gates. The circuit diagrams for the two input C-SET NAND and NOR gate are shown in Fig. 6.6(a) and Fig. 6.6(b), respectively. Each gate uses four SETs, and the input voltages V_A and V_B are applied to the gate terminals of the SETs. The basic circuits are very similar to CMOS NAND and NOR gates.

We consider first the C-SET NAND gate (Fig. 6.6[a]). The 'truth table' for the gate is shown in the inset to Fig. 6.6(a). SET1 and SET2, connected in parallel between the output voltage V_o and the supply voltage V_{dd}, are biased by a trimming voltage V_{t1} (not shown) in a manner similar to the situation shown in Fig. 6.2(b) for the C-SET inverter. These devices then operate analogous to p-MOSFETs and turn 'on' for 'low' input voltage. SET3 and SET4, connected in series between the output voltage V_o and ground, are biased by a different trimming voltage V_{t2} (not shown) in a manner similar to the situation shown in Fig. 6.2(c). These devices then operate analogous to n-MOSFETs and turn 'on' for 'high' input voltage. V_o is pulled 'low' only if both the SET2 and SET3 are 'on', i.e. V_A and V_B are both 'high'. For this condition, SET1 and SET2 are 'off' and there is a high resistance between the output terminal and V_{dd}. If either or both SET1 and SET2 are 'off', i.e. either or both V_A and V_B are 'low', at least one SET between the output and ground is 'off' and there is a high resistance between these terminals. In contrast, at least one SET between the output and V_{dd} is 'on', pulling V_o 'high'. The circuit then operates as a NAND gate.

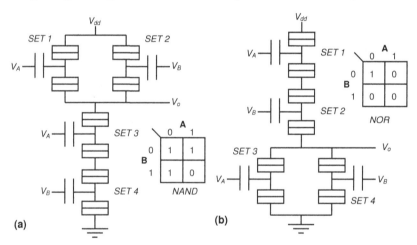

Fig. 6.6 Complementary SET gates. (a) NAND gate. (b) NOR gate.

The C-SET NOR gate is shown in Fig. 6.6(b), with the 'truth table' for the gate shown in the inset. Here, SET1 and SET2, operating analogous to p-MOSFETs, are connected in series between the output voltage V_o and the supply voltage V_{dd}. SET3 and SET4, operating analogous to n-MOSFETs, are connected in parallel between V_o and ground. V_o is pulled 'high' only if both the SET1 and SET2 are 'on', i.e. V_A and V_B are both 'low'. For this condition, SET2 and SET3 are 'off' and there is a high resistance between the output terminal and ground. If either or both V_A and V_B are 'high', then the output is pulled 'low'. The circuit then operates as a NOR gate.

Differences in the transfer characteristics of SETs and MOSFETs lead to differences in the operation of the C-SET gates compared to CMOS gates. For example, the SETs are 'on' in a narrower range of gate voltage in comparison with MOSFETs, and reliable operation of the gates may require a quasi-static switching technique (Tucker, 1992). It may, however, be possible to improve this situation by slight modifications of the basic circuits. Chen *et al.* (Chen *et al.*, 1996) have proposed modified inverter, NOR and NAND gates where the ground terminal is replaced by $-V_{dd}$, and the SETs are arranged symmetrically between the output voltage and the $+V_{dd}$ and $-V_{dd}$ terminals. These gates have symmetrical 'high' and 'low' voltage levels with reference to ground, and operate better.

Theoretical analysis of the performance of the devices (Chen *et al.*, 1996) suggests that the maximum operating temperature T_{max} ~$0.025e^2/Ck_B$. For a load capacitance C_o ~ 1 fF, the switching speed was ~0.7 ns and the power consumption per transistor P ~10^{-9} W. The switching speed for these devices is not very high, although this may be raised in a densely packed circuit where the load capacitance is reduced by minimizing interconnect capacitances. This would also raise the device density compared to CMOS even further. The power consumption also appears to not be very low, and at a device density of $10^{11}/cm^2$, would correspond to 100 W/cm^2 if all transistors were activated.

Regarding power consumption, charge-state single-electron logic, where the presence or absence of an electron can define a bit, may have better performance. Chen *et al.* also note that, while the gates appear to be very sensitive to offset charge variation, requiring this to be only ~0.03e, this variation may reduce inherently for very small dimensions ~1 nm. The use of multiple-tunnel junction (MTJ) SETs may also reduce sensitivity to offset charge. In addition, the maximum operating temperature can also be higher with MTJs, e.g. for a 5-junction transistor, theoretical simulation suggested a 2.5 x increase in the maximum operating temperature for the same feature size (Chen and Likharev, 1998).

From the preceding discussion, it is clear that there are considerable difficulties in a practical realization of C-SET logic gates. However, a C-SET NAND gate similar to the circuit of Fig. 6.6(a) has been fabricated using silicon nanowire SETs by Stone and Ahmed (Stone and Ahmed, 1999). The silicon nanowires, defined in the 40 nm-thick top silicon layer of SOI material, were approximately 30 nm in width and 300 nm in length, and formed MTJs at 4.2 K. The NAND gate was operated at 1.6 K, with an output voltage swing of ~2 mV. While the operating temperature and the output swing were very low and the voltage gain was less than unity, the device demonstrates that a direct implementation of the C-SET NAND gate is at least possible. It remains to be seen if improvements in the performance of the device can lead to circuit application.

6.2.4 Programmable SET logic

Uchida *et al.* (Uchida *et al.*, 2003) have proposed a programmable logic architecture for SETs, where a SET integrated with a charge storage node is used to implement both n-MOS- and p-MOS-like devices. A circuit diagram of the device is shown in Fig. 6.7(a). The SET island is coupled directly to the gate via the capacitance C_{gl}, and to a storage or memory node via the capacitance C_{sl}. The storage node also couples to the gate, via the capacitance C_{gs}. Depending on the presence of charge on the storage node, the SET operates as an n-MOS- or p-MOS-like, device. Without any charge on the storage node, for the voltage range V_{g1}–V_{g2}, the SET operates as an n-MOS-like device. However, if the gate voltage V_g is swept to a higher value V_{g3}, charging of the storage node occurs, biasing the SET island and shifting the single-electron oscillations along the gate voltage axis. The voltage V_{g3} is chosen such that a phase change of π occurs in the oscillations, and for the voltage range V_{g1}–V_{g2}, the SET operates as a p-MOS-like device. The writing of a charge onto the storage node can then be used to programme the SET to form either an n-MOS- or p-MOS-like device

Figure 6.7(b) shows the circuit diagrams for the inverter, NOR gate and NAND gate. All the devices use a resistive load R for 'pull-up' of the output. The inverter requires only one n-MOS-like SET, used as a 'pull-down' device. Programming the SET as a *p*-MOS device converts the circuit into a buffer. The NOR gate can be implemented using two *n*-MOS-like SETs, and the AND gate using two *p*-MOS-like SETs. The basic circuit of the NOR and AND gate is the same and can be programmed to implement either function by writing to the SET storage nodes. The use of 'pull-down' devices in parallel helps in overcoming the high resistance of SETs and improves switching speed. Uchida *et al.* also propose a programmable logic array (PLA) using these devices, to implement arbitrary logic functions in a structured manner. The various devices in the array can be programmed to behave either as *n*-MOS- or *p*-MOS-like devices, as required, to implement a given logic function.

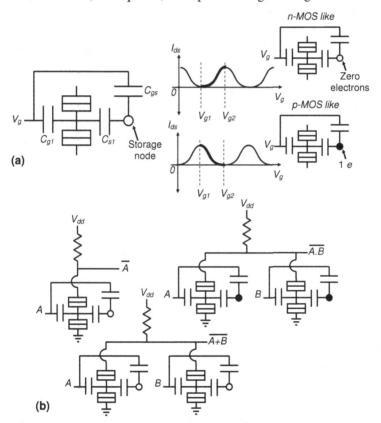

Fig. 6.7 Programmable SET gates (after Uchida *et al.*, 2003). (a) Circuit diagram of basic device, and biasing relative to single-electron oscillations. (b) Circuit diagram for inverter, NOR and NAND gates.

Uchida et al. have demonstrated the operation of the programmable buffer/inverter at room temperature, using SETs fabricated in an ultra-thin film of silicon (Uchida et al., 2001). Here, the top silicon layer of the SOI material was thinned to only a few nanometers. Surface roughness combined with quantum-confinement effects caused fluctuations in the potential profile along the film, isolating regions along the film. The device behaved as a combined SET/MOSFET, and as the gate voltage was increased, a narrow channel was formed between the source and drain along potential valleys. Nearby valleys, isolated from the channel, formed charge storage nodes. Increasing the gate voltage could trap charge in these nodes, which biased the channel potential.

Uchida et al. observed very clear current peaks in the device, with a maximum peak-to-valley ratio of ~100. The position of the peaks could be shifted by π by applying ~8 V on the gate. The output voltage swing at 300 K was ~5 mV, and an integrated CMOS inverter was used to amplify this to ~0.1 V. Alternative designs of programmable SETs may also be possible, e.g. Saitoh et al. (Saitoh et al., 2005) have proposed a programmable SET using a floating gate consisting of silicon nanocrystals, deposited on top of a single-hole transistor. Here, the storage node is formed by the silicon nanocrystals and is remote from the thin film forming the device channel.

Uchida et al. (Uchida et al., 1999) have also proposed a logic architecture using a 'pre-charge' and 'evaluation' phase. The programmable logic devices discussed above can be utilized in this scheme as well. The architecture uses the SETs only as 'pull-down' devices, i.e. the SETs are used only to discharge an output capacitor. The use of a low supply voltage is proposed, which helps in obtaining complete discharge of the output capacitor without SETs turning 'off'. A 'pre-charge' phase can be used to charge the load capacitor fully to the supply voltage, with the 'pull-down' SETs switched out of the circuit. This can then be followed by an 'evaluation' phase, where the SET logic array is switched into the circuit and the load capacitor charged or discharged depending on the logic inputs. A CMOS amplifier can be used to amplify the relatively low voltages across the output capacitor.

6.2.5 Logic using SETs with multiple input terminals

It is possible to construct a two input XOR gate using a single SET with multiple input gate terminals (Takahashi et al., 2000b). Here, the output is '1' only if one of the input terminals is '1'. Such a device utilizes the oscillating nature of the SET I_{ds}-V_{gs} characteristics to turn the device 'on' at two different points, corresponding to two combinations of the input gate voltages. A circuit diagram of the device is shown in Fig. 6.8(a). The circuit uses a resistive load such that if the SET is 'on', the output capacitor C_o is discharged and the output V_o is pulled down. The SET island is capacitively coupled to two separate gate terminals, and the two input voltages V_A and V_B are applied to these terminals. For simplicity, we assume that the gate capacitance for each terminal is the same. The total gate voltage for the SET is therefore the sum of the two input voltages, $V_g = V_A + V_B$ (Fig. 6.8[b]).

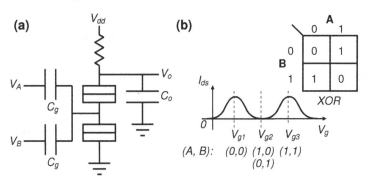

Fig. 6.8 Two-input XOR gate using a dual gate SET. (a) Circuit diagram. (b) Biasing relative to single-electron oscillations, shown schematically.

The SET is biased such that for zero voltage applied to V_A and V_B, i.e. for the input combination (0, 0), the SET operates at a Coulomb oscillation peak, at V_{g1}. The biasing of the SET at V_{g1} can be arranged either by using an additional gate terminal, or by applying a constant offset voltage to one of the two input gate terminals. For the input combination (0, 0), the Coulomb blockade is overcome, the SET is 'on' and V_o is pulled down to a low voltage, i.e. output '0'. For the input voltage combination (1, 0) or (0, 1), the magnitude of V_A and V_B is adjusted such that the SET is biased at the next Coulomb oscillation valley at V_{g2}. This implies that the input voltage corresponding to a '1' is $V_A = V_B = V_{g2} - V_{g1}$. For input (1, 0) or (0, 1), the SET is then 'off' and V_o is pulled up via the load resistor towards V_{dd}, i.e. the output is '1'.

Finally, if both V_A and V_B are '1', then the SET is biased at a gate voltage $V_{g3} = 2(V_{g2} - V_{g1})$, corresponding to the next oscillation peak. The SET is 'on', pulling V_o low, i.e. the output is '0'. The circuit then implements an XOR gate, with the truth table shown in the inset to Fig. 6.8(b). The circuit may be extended to three inputs with an additional gate capacitor. Again, the sum of all the gate voltages should bias the SET at an oscillation peak only if all inputs are 'low' or 'high'. For intermediate voltages, the gate capacitances are such that the SET is not biased at an oscillation peak, though it may not be possible to arrange the gate capacitances such that the SET turns 'off' for all input combinations. This would prevent complete pull-up of the output and reduce the voltage swing.

Various SET designs have been used to implement the XOR gate and the exclusive-not-OR gate, and room temperature operation has been demonstrated (Takahashi et al., 2000b; Saitoh et al., 2003; Kitade and Nakajima, 2004; Kitade et al., 2005). Takahashi et al. (Takahashi et al., 2000b) implemented an XOR gate operating at 40 K using a multiple-gate SET defined by PADOX. Two different designs of XOR gates were demonstrated, the first with two top gates arranged in the same plane above the island, and the second with the two gates stacked one above the other. The first configuration had approximately equal capacitances to the island, and the XOR gate operated in a manner similar to the discussion above (Fig. 6.8).

The second configuration raised the possibility of voltage gain and a smaller island capacitance, but led to different gate capacitances. This implied that V_A and V_B had different levels of gating and for similar values of V_A and V_B, the SET would not be biased at the optimum position at the top of a conductance peak. The smaller capacitance of this structure increased the peak-to-valley ratio to an extent that the output voltage on-off ratio was larger than in the SET with in-plane top gates. Room temperature operation of the XOR gate has also been demonstrated, using a single-hole transistor (Saitoh et al., 2003), and using a MTJ SET (Kitade et al., 2005). These devices have very small islands, ~10 nm or less in size, and large peak-to-valley ratios, e.g. the device of Kitade *et al.* (Kitade *et al.*, 2005) has a maximum peak-to-valley ratio of 77 at 300 K. It may also be possible to generalize the concepts of using multiple-gate SETs to fabricate other logic gates. Kaizawa *et al.* (Kaizawa *et al.*, 2006) have theoretically proposed a device with an array of charging islands controlled by multiple gates. By using some of the gates as input terminals, and others as control gates, various logic functions such as XOR, AND and OR may be implemented using the same device.

A majority logic architecture has also been proposed using multiple-gate SETs (Iwamura *et al.*, 1998). In such an architecture, each logic gate has an odd number of input terminals and the output is determined by the state of the majority of the input terminals, i.e. if the majority of inputs are '1', then the output is also '1'. Each input terminal can be capacitively coupled to the island of a SET, forming a multiple-gate SET with an odd number of gates. The gate capacitances may be arranged such that for a low voltage on the majority of gates, the output is low, and for a high voltage on the majority of gates, the output is high. However, the design of such a gate is complicated by the oscillatory nature of the SET transfer characteristic, requiring careful design of the gate capacitances. The multiple-gate terminals also tended to raise the total capacitance of the SET, leading to a lower operating temperature.

6.2.6 Effect of offset charge

A general problem in the implementation of all SET-based logic gates, whether resistively-loaded, complementary or multiple-gate designs, is the high sensitivity of the constituent SETs to switching due to random offset charges. Fractional variation in the offset charge can switch the SET from 'on' to 'off', leading to errors in the computation.

In the experimental devices discussed above, additional gates are used to tune offset charges out from device to device, and to bias the SET at the required operating point. Clearly, very good control over offset charge is necessary to realize a practical LSI circuit. Likharev (Likharev, 1999) has pointed out that, even with a low trap/defect concentration of only $10^{10}/cm^2$, 1 in 1000 devices fabricated with 1 nm islands would suffer from background charge fluctuation. This would strongly influence any LSI circuit application, and it may be necessary to incorporate error correction in the architecture.

However, the possibility exists that, for islands of very small dimensions ~1 nm, a 'self cleaning' process may lead to a reduction in the number of defect states per island, reducing the impact of the problem (Chen *et al.*, 1996; Likharev, 1999). A further possibility is the use of offset-charge independent logic architecture. Such an architecture is possible using resistively coupled SETs (Likharev, 1987), which are offset-charge insensitive.

The novel single-electron logic proposals of Kiehl and Ohshima (Kiehl and Ohshima, 1995; Ohshima, 1996; Ohshima and Kiehl, 1996), where the logic states are defined by the phase of single-electron tunnelling oscillations in *single* tunnel junctions, also appear to be insensitive to offset charge. However, the general difficulties in observing single-electron tunnelling oscillations (Chapter 2) may prevent the realization of such a device.

6.3 Charge State Logic

Charge state single-electron logic circuits use the presence or absence of a small charge packet consisting of a single electron, or at most a few electrons, to represent a bit. This approach explicitly uses the ability of a single-electron device to precisely control charge. A more robust implementation for a practical circuit may, however, require the bits to be represented by a few electrons rather than only one electron. Charge state logic circuits can be designed using a variety of different approaches, all of which use the presence or absence of the bit at a specific point in the circuit to represent the output.

Perhaps the most well-developed of these approaches is binary decision diagram (BDD) single-electron logic (Asahi *et al.*, 1995, 1997), where the logic function is implemented using a network of two-way switching nodes. Here, depending on the input logic values, an electron is transferred through the network into one of two output branches, representing the '1' or '0' outcome. Another approach is the quantum cellular automaton (QCA), mainly investigated theoretically (Lent et al., 1993a, 1993b, Tougaw et al., 1993). A QCA is formed by an array of cells, where each cell consists of an arrangement of quantum dots (QDs) charged by two electrons. The direction of the charge polarization of a cell represents the '1' or '0' bit and the polarization of a given cell can switch the polarization of a neighbouring cell.

Depending on the state of a set of input cells, the QCA relaxes into a 'ground state', where another set of cells represent the output. The QCA approach does not require interconnects, i.e. it is 'wireless', and theoretically has the potential for fast switching and low power dissipation. Alternative wireless approaches are also possible, such as the single-electron 'parametron' array, where the cells are formed by three QDs, and are switched using an external AC electric field (Korotkov and Likharev, 1998).

A wireless single-electron logic architecture has been proposed where a short chain of islands defines a cell (Korotkov, 1995). Again, the bits are represented by the polarization of the cell, and the cells may be switched using an AC electric field. Alternative versions of cellular circuit architecture may use local interconnects between cells – a cellular architecture using interconnected electron pumps has been proposed (Ancona and Rendell, 1995; Ancona, 1996). This architecture can directly simulate specific problems, e.g. the simulation of a 'lattice gas'. AND, OR and XOR gates, and memory cells, based on interconnected electron pumps have been proposed as the building blocks for this architecture (Ancona, 1996).

In the following, we discuss the BDD, QCA and single-electron parametron approach to charge-state single-electron logic in detail. An emphasis is placed on BDD logic, at present the most well developed of these logic proposals. In this regard, the fabrication and characterization of a basic BDD logic element, the two-way switch of He et al. (He et al., 2004a, 2004b), will be discussed in detail.

6.3.1 Binary decision diagram logic

Binary decision diagram logic architecture (Asahi et al., 1995, 1997) provides a new paradigm for logic systems, particularly suited for single-electron and quantum effect devices. Conventional logic architectures have been optimized over a large number of years to operate with existing, CMOS-type devices. This implies that it is difficult for single-electron devices, with very different strengths and weaknesses, to compete with existing CMOS devices using conventional logic architectures. Binary decision diagram logic architectures can be designed around the inherent advantages of single-electron devices, e.g. precise control of charge and low power consumption. The architecture may also be suitable for other novel devices, e.g. optical switching and electron-wave modulation devices (Asahi et al., 1995, 1996). Here, we concentrate on single-electron implementations of BDD logic.

The conventional approach to logic design represents a digital function by a Boolean equation, and then implements this equation using a series of logic gates defined using switching transistors. This requires high gain in the transistors, necessary in order to obtain sharp turn-on or turn-off characteristics, voltage matching of the output levels, and adequate fan-out and fan-in. The high switching performance of the MOSFET is particularly suited for such an implementation. However, single-electron devices have poor voltage gain and do not form very good switching elements. This complicates the implementation of a CMOS-type logic architecture using SETs.

In contrast to CMOS logic, the BDD or 'pass transistor' logic approach, originally proposed as a tool for computer-aided logic design (Akers et al., 1978, Bryant et al., 1986), represents a digital function using a directed graphical representation. The entire digital function is represented as a tree-like diagram, constructed using two-way switching elements. A hardware implementation of BDD logic using CMOS, investigated by Yano et al. (Yano et al., 1996b), suggested that circuit area, delay and power dissipation were improved and smaller device counts were possible in comparison to conventional circuits.

Concurrently, Asahi et al. (Asahi et al., 1995, 1997) proposed the implementation of BDD logic by SETs, and by other quantum devices. In such an implementation, large voltage gain, large device drivability, and voltage level matching are not necessary. The current through the circuit can be very small and single-electron transfer is possible, implying very low power dissipation.

6.3.1.1 Binary decision diagram logic: Basic logic gates

A BDD logic circuit consists of a series of two-way switching elements or 'nodes', connected to form a circuit tree (Fig. 6.9[a]). A single-input branch leads into the circuit tree. The output of the circuit consists of two branches, or 'leaves', the '0' output branch and the '1' output branch. The circuit operates by injecting a charge packet or 'messenger' into the input branch. The messenger can consist of a single electron (shown), or a few electrons. A series of switching signals X_i at each node i, transfer the messenger electron into one of the two output branches of the node. The BDD circuit tree is designed such that it directly implements the logic function and, depending on the input variables X_i, the messenger electron arrives either at the '0' output branch, or the '1' output branch. Which branch the messenger arrives in constitutes the logic function output, e.g. if the electron arrives at the '1' output branch, then this corresponds to an answer of '1'.

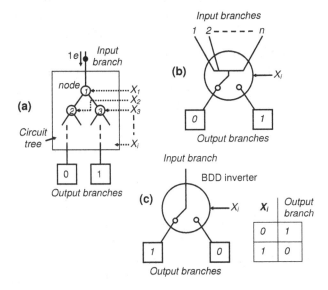

Fig. 6.9 BDD logic. (a) BDD logic circuit tree. (b) BDD logic n-input switching node. (c) BDD logic inverter.

Figure 6.9(b) shows the basic two-way switching node. Here, 'n' input branches lead into the node, and two output branches lead out of the node. Typically, in BDD implementations of simple, two input logic functions, only one or two branches lead into a node. The switching variable X_i transfers a messenger electron from the input branches into one of the two output branches, e.g. if the switching variable is 'high', $X_i = 1$, then the '1' output branch is selected. Exchanging the labels of the '1' and '0' output branches converts the basic switching node with one input into a BDD NOT gate or inverter (Fig. 6.9[c]), e.g. $X_i = 1$ leads to the messenger electron transferred into the '0' output branch.

Figure 6.10 shows the BDD circuits for two-input NAND, NOR and XOR gates. To understand the operation of the NAND gate (Fig. 6.10[a]), we may trace the path of the messenger electron through the gate for different input combinations. The outputs corresponding to the various input combinations are shown in the truth table. For example, when inputs (X_1, X_2) correspond to (1, 1), the messenger is guided from branch *a*, to *b*, to *e*, to the '0' output branch. Alternatively, for input (1, 0), the messenger is guided from branch *a*, to *b*, to *d*, to the '1' output branch. Similarly, other input combinations also produce the correct output. We note that the tree is asymmetric, and the number of branches traversed for different input combinations is different, e.g. if $X_1 = 0$, only branch *c* is used. Two branches, *c* and *d*, feed into the '1' output branch, but only one branch, *e* feeds into the '0' output branch. As we will see below, this asymmetry may require the introduction of 'dummy' nodes in a multiple-stage BDD circuit. We also note that by redefining the '0' output as the '1' output and vice versa, the circuit implements the AND function.

Figure 6.10(b) shows the BDD circuit for the NOR gate. Again, the circuit is asymmetric, and again, by exchanging the labels of the output branches, the circuit implements the OR function. In addition, by inspecting the connections of the tree, it can be seen that the circuit forms the NAND gate, if inverted input values $\bar{X}_i$ are applied.

Finally, Fig. 6.10(c) shows the BDD circuit for the XOR gate. Unlike the NAND and NOR gates, the XOR gate circuit is symmetric and for all combinations of the input variables, the messenger electron traverses two branches to reach the required output node. More complex circuits, such as a 4-bit adder and a 4-bit comparator, have also been proposed and simulated (Asahi *et al.*, 1998). In addition, for digital systems with closely related functions, part of the BDD circuit for two or more functions may be shared. This is referred to as 'shared BDD' logic (Yamada et al., 2001) and allows a reduction in the number of BDD nodes.

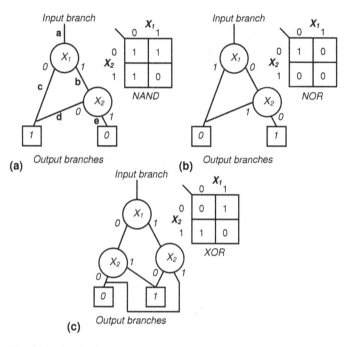

Fig. 6.10 BDD logic gates. (a) NAND gate. (b) NOR gate. (c) XOR gate.

Asahi et al. (Asahi et al., 1995, 1997) proposed that SETs provide a very suitable means to implement the nodes in a BDD circuit, as the circuit does not require gain. The BDD circuit simply switches the messenger electron into the required output branch and there is no requirement for gain in the nodes constituting the circuit, a situation very different from conventional CMOS logic. In CMOS logic, a Boolean expression is implemented by a combination of basic logic gates, the input voltage level to a first stage of logic gates switching the output voltage of these gates, which is then used to drive a further stage of gates. The requirement that a stage can drive other stages, i.e. 'fan-out', requires high gain transistors for the gates. SETs, with their very low (often lower than unity) gain, are clearly less suited to implementing this scheme compared to conventional MOSFETs. In contrast, gain is unnecessary for BDD circuit nodes and SETs are suitable devices for their implementation.

Figure 6.11(a) shows the circuit diagram for a SET implementation of the BDD node. The node uses two SETs with the source terminals connected to the input branch. The drain terminals of the SETs form the output branches '1' and '0'. The input variable X_i, and its converse value $\bar{X}_i$, are applied to the gates of the SETs, turning one of the SETs 'on' and the other 'off'. This is arranged by biasing the SET such that for a low gate voltage, the SET is biased between two gate oscillation current peaks, i.e. within the Coulomb blockade region, and for a high gate voltage, it is biased at a current peak. Therefore, if $X_i = 1$, SET1 is 'on' and SET2 is 'off', and vice versa for $X_i = 0$ (Fig. 6.11[a], insets). Two clock signals φ_0 and φ_1, capacitively coupled to the source and drain of the SETs, are used to drive the circuit and transfer a messenger electron from the input to an output branch, in a controlled, timed manner.

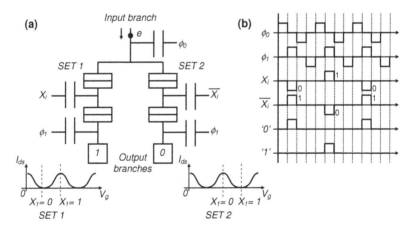

Fig. 6.11 SET implementation of a BDD logic node. (a) Circuit diagram. (b) Timing diagram.

The timing diagram of Fig. 6.11(b) illustrates the operation of the circuit. The messenger electron is transferred from the input branch to one of the two output branches when a positive pulse is applied to φ_1, and simultaneously, a pulse is applied to X_i. For example, if $X_i = 1$, then SET1 is 'on' and SET2 is 'off'. A positive pulse on φ_1 then creates a positive bias at the drain terminal of the SETs, pulling the messenger electron from the input branch, through SET1, into output branch 1.

Similarly, if $X_i = 0$, then SET2 is 'on' and SET1 is 'off, and the electron is transferred through SET2 to output branch 0. The upper clock φ_0 is not directly used in the electron transfer here. However, if preceding stages exist with the node, a positive pulse on φ_0 draws an electron into the input branch. We note that φ_0 and φ_1 are $\pi/4$ out of phase with each other, such that if one of the clocks is pulsed to positive or negative voltage, the other is at zero voltage. This implies that when an electron is transferred through a SET, the source voltage is zero and the drain voltage is proportional to the pulse magnitude. The scheme can be extended to a four-phase clocking scheme (Asahi *et al.*, 1997). As noted earlier, if we exchange the labels of the two output branches, the circuit forms a BDD NOT gate.

Figure 6.12(a) shows the circuit diagram for a BDD NAND gate, defined using four SETs. Three clock signals φ_0, φ_1 and φ_2 are used to drive the circuit. In addition, a 'dummy' timing stage is used after SET1, to create a consistent timing scheme with equal numbers of timed stages along any path from the input to the output branches. As we shall see below, no matter what the path, the messenger electron then arrives at an output branch when φ_2 becomes positive, after $\pi/2$ of a cycle.

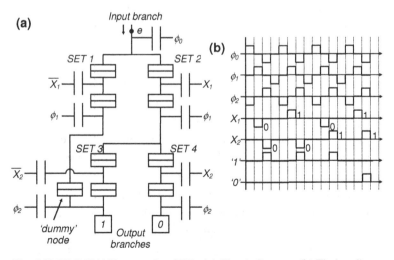

Fig. 6.12 BDD NAND gate using SETs. (a) Circuit diagram. (b) Timing diagram.

Figure 6.12(b) shows the timing diagram for the circuit for all combinations of X_1 and X_2. In a manner similar to the general BDD node shown in Fig. 6.11, each successive clock signal is $\pi/4$ out of phase with the previous one. A messenger electron is injected into the circuit from the input branch when φ_0 is positive. Depending on the input variable X_1, when φ_1 becomes positive after $\pi/4$ of a cycle, the electron is transferred either to node 'a' or 'b'. Finally, when φ_2 becomes positive, depending on the value of X_2, the electron is transferred to one of the two output branches. It takes $\pi/2$ of a cycle for the electron to arrive at an output branch. For example, if $(X_1, X_2) = (1, 0)$, the electron transfers via SET2 and SET3 to output '0'. In contrast, if $X_1 = 0$, then no matter what the value of X_2, the output is zero. In order to prevent the electron arriving at output '0' after only $\pi/4$ of a cycle, a dummy timed stage is used, and the electron only crosses this after a further $\pi/4$ of a cycle.

We note that the NAND gate requires only three clock stages. However, the timing scheme as shown can accept a fourth clock signal and a further timed stage. For gates requiring even more stages, the first clock can be reintroduced into the circuit, starting the four-phase clocking sequence again. Asahi *et al.* have also proposed and simulated circuits for a BDD XOR gate (Asahi *et al.*, 1997) and for a BDD 4-bit adder and comparator (Asahi *et al.*, 1998). The latter two circuits are more complex and require a total of 12 timed stages. A four-phase clock, introduced into the circuit three times, is necessary to drive the circuit.

Asahi *et al.* (Asahi *et al.*, 1997, 1998) have also proposed a general architecture for a logic system using BDD circuit trees, shown in Fig. 6.13. The BDD devices are built into a cascade to form the BDD circuit tree, driven using a four-phase clock. An electron is injected into the tree and switched by the n input variables X_n to the correct output branch after a number of stages of the clock. The electron can then either be fed back into the tree via a feedback loop (shown by the dotted line) for successive logic operations, or a new electron can be injected for each logic operation. The presence of the electron on either the '1' or '0' output branch can be detected using a single-electron latch circuit (Asahi *et al.*, 1998), producing a voltage output signal.

Asahi *et al.* (Asahi *et al.*, 1998) have also investigated the operation error for a 4-bit adder, caused by unwanted electron tunnelling. The error mechanisms in these systems can include stochastic fluctuations in the electron number, errors in the selection of branches and thermal fluctuations of electrons. For relatively large 10 aF tunnel capacitances, and 100 kΩ tunnel resistances, they found that the error rates were somewhat high, $\sim 10^{-2}$–10^{-1} at 1 K. However, recent room-temperature-operating SETs have capacitances well below 1 aF, and better performance may be possible.

Alternatively, a majority decision over a number of logic trials may be used to decide the output. Yamada *et al.* (Yamada *et al.*, 2001) have simulated the performance of a 2-bit BDD adder, where the tunnel junction capacitances were 10 aF, the island stray capacitance was 20 aF, and the tunnel resistances were 1 MΩ with the SET in the 'on' state. By using multiple counts of the arrival of electrons at the output branches, the circuit could operate at a higher temperature of 20 K. Again, the simulation capacitances here are rather high, and with better SETs, room-temperature operation may be possible.

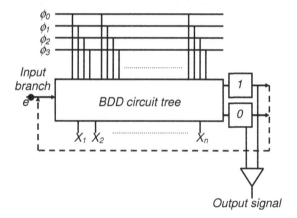

Fig. 6.13 BDD circuit architecture (after Asahi *et al.*, 1998).

6.3.1.2 Implementation of BDD logic circuits in GaAs

In early experimental work on BDD logic circuits, Tsukagoshi *et al.* fabricated and demonstrated the operation of the basic BDD node, and the BDD AND gate, using electron turnstiles based on GaAs nanowire SETs (Tsukagoshi *et al.*, 1998; Tsukagoshi and Nakazato, 1998). While these devices are fabricated in GaAs and are not compatible with LSI processes, we will discuss them here as the ideas they develop may be extended to devices in Si. The devices are similar to, and compatible with, the technology developed for the first single-electron memory circuits (Nakazato et al., 1994), and may be regarded as providing a logic system complementary to these circuits. In more recent work on BDD devices, a node, AND, OR and XOR gates have been fabricated using a close-packed hexagonal network of GaAs SETs (Hasegawa and Kasai, 2001; Kasai and Hasegawa, 2002; Nakajima et al., 2002b; Nakajima et al., 2003; Miyoshi et al., 2005).

Binary decision diagram logic architecture, where the same basic element is used to construct the logic circuit, is particularly suited for circuit implementation using an ordered array of devices. While this implementation of BDD circuits using hexagonal networks is in GaAs as well, similar circuit layouts may be possible in silicon. However, circuits in GaAs can lend themselves to the self-assembly of the SET islands using growth techniques (Nakajima et al., 2002b), allowing control over the island size and reducing device–to-device variation in the circuit. A full self-assembly process is difficult in crystalline Si, but may be possible using silicon nanocrystal films.

Tsukagoshi and Nakazato (Tsukagoshi and Nakazato, 1998) fabricated and demonstrated the basic BDD node or two-way switch using few-electron turnstiles fabricated in δ-doped GaAs material. The turnstiles, based on earlier work on few-electron pumps and turnstiles in GaAs (Tsukagoshi et al., 1997; Tsukagoshi and Nakazato, 1997), were able to transfer charge packets ~100 electrons in size. The charge packet size is ~500–1000 times smaller than conventional CMOS devices.

While the turnstiles are not *single*-electron devices, in the sense that they do not transfer charge packets one electron in size, the turnstiles do rely on SETs to operate and are Coulomb blockade *controlled*. Figure 6.14(a) shows a schematic of the device. It consists of three SETs joined at a central node. Each SET is formed by a nanowire defined in the δ-doped GaAs. Disorder in the doping along the nanowire defines a MTJ. In-plane side-gates are used to bias the MTJ and control the Coulomb blockade. A supply voltage V_{in} biases the input branch and drives a current into the circuit (Fig. 6.14[b]).

A radio frequency (r.f.) signal V_{rf}, capacitively coupled to the central node, forms the clock and transfers charge packets along the turnstile formed by SET1 and SET2, or SET1 and SET3, i.e. the current is switched into the output '1 ' or output '0' branch (For details of turnstile operation, see Chapter 5). The input variable to the circuit X_I is applied as the gate voltage V_g, of similar magnitude but opposite polarity, to SET2 and SET3. This biases the two SETs in a manner such that for positive V_g, SET2 is 'on' and SET3 is 'off', and for negative V_g, SET3 is 'on' and SET2 is 'off'.

Figure 6.14(c) schematically shows I_0 and I_1 as a function of the frequency f of $V_{r.f.}$. Applying a positive or negative value of V_g then switches current into branch 1 or branch 0. The current increases linearly with f, as expected in a turnstile where the current $I = nef$, for a charge packet n electrons in size. Tsukagoshi and Nakazato (Tsukagoshi and Nakazato, 1998) operated their circuit at 4.2 K, using $V_{in} = 5$ mV, $f = 0.5$–6 MHz, and $V_g = 0.5$ V. Switching action could be observed even if the input r.f. power was reduced to -20 dBm, corresponding to the switching of a packet of ~100 electrons through the circuit.

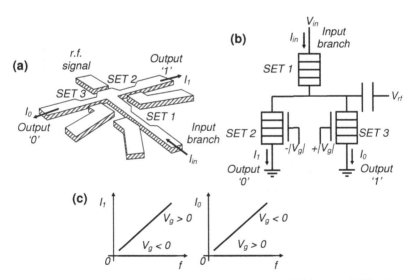

Fig. 6.14 BDD logic two-way switch (after Tsukagoshi and Nakazato, 1998). (a) Schematic diagram of device. (b) Circuit diagram. (c) Output currents vs. clock signal frequency f.

Tsukagoshi et al. (Tsukagoshi et al., 1998) used their GaAs BDD node to fabricate a BDD AND gate, modified such that the SETs operated as few-electron pumps rather than turnstiles, i.e. without a directly applied DC bias (for details of electron pump operation, see Chapter 5). In addition, the δ-doped GaAs SETs used Ti/Al Schottky gates rather than side-gates. The circuit diagram of the device is shown in Fig. 6.15(a). The device uses eight GaAs nanowire SETs, arranged to form a series of few-electron pumps. SET1, SET2 and SET3 form the BDD node X_1, with SET2 leading to the '1' output path for the node, and SET3 leading to the '0' output path for the node. The input variable X_1 is applied to the gates of SET2 and SET3, such that if $X_1 = 1$, SET2 is 'on' and SET3 is 'off', selecting the '1' output path. Conversely, $X_1 = 0$ selects SET3 and the '0' output path. In a similar manner, SET3, SET4 and SET5 form the BDD node X_2. The remaining SETs are used to complete the electron pumps and allow an equal number of clocking stages along all possible current paths. The full circuit forms the BDD AND gate shown in Fig. 6.15(b). The SETs are arranged somewhat differently from the NAND circuit (Fig. 6.12) discussed in the previous section.

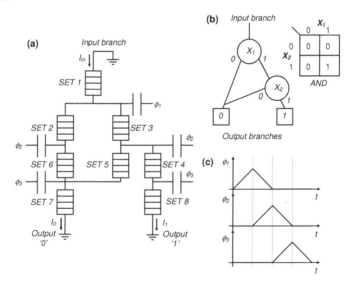

Fig. 6.15 BDD logic AND gate (after Tsukagoshi et al., 1998). (a) Circuit diagram. (b) Binary decision diagram. (c) Three-phase clock signals.

A three-phase capacitively coupled clock, with amplitudes ϕ_1, ϕ_2, and ϕ_3, is used to drive the circuit. The pulses applied to ϕ_1, ϕ_2 and ϕ_3 are $\pi/2$ out-of-phase with the previous pulse. As these pulses are applied, the Coulomb blockade in the SETs is overcome sequentially. This transfers a messenger electron packet through the circuit, along one of three current paths, to either the output '1' or output '0' branch, assuming the SETs along the required current path are 'on'. The three-phase clock is similar to the proposals of Asahi *et al.* discussed in the previous section, (Asahi *et al.*, 1997), where a multi-phase clock is used to drive the messenger through the BDD circuit.

In a manner similar to their two-way switch, Tsukagoshi *et al.* arranged the SETs in pairs such that if one of the SETs was 'on', the other was 'off'. These pairs were formed by SET2 and SET3, and by SET5 and SET6. The input variables X_1 and X_2 were then used to switch the pairs, e.g. if $X_1 = 1$, and $X_2 = 0$, SET2 was 'off', SET3 was 'on', SET5 was 'on', and SET6 was 'off'. This transferred the messenger electron packet into the '0' output branch. Similarly, other combinations of X_1 and X_2 also switched the messenger into the required output branch. Tsukagoshi *et al.* operated their circuit at 1.8 K, using clocking pulses with amplitude of 25 mV peak-to-peak, and frequency of 1 MHz. A charge packet of 160 electrons could be switched through the circuit.

Binary decision diagram logic circuit trees, constructed using the same basic node and branch elements, inherently lend themselves to implementation using ordered device layouts. Hasegawa and Kasai (Hasegawa and Kasai, 2001; Kasai and Hasegawa, 2002) have proposed the use of hexagonal close-packed arrays of devices for a systematic implementation of BDD circuits (Fig. 6.16). Such an array inherently allows for a high density of devices. The basic BDD node in this scheme is implemented using a Y-shaped two-way switch.

Hasegawa and Kasai propose two types of switches, a 'node' switch and a 'branch' switch (Fig. 6.16[a]). Both types use three nanowires, arranged in a Y-shaped layout. One of the nanowires forms the input branch, and the other two form the '0' and '1' output branches. The 'node switch' uses a gate electrode on each branch to create tunnel barriers, with an island formed at the centre of the switch. The nanowires can form quantum point-contacts, where the conductance is quantized in multiples of $2e^2/h$.

The Schottky gates can be used to turn a branch 'off' or 'on', and switch current through the device by closing the electron channels completely or allowing an electron channel to open. In addition, single-electron charging effects can occur in the central island and, in this case, the input branch, combined with one of the output branches, forms a SET. The various gate voltages, which control not only the tunnel barriers but also the island potential, may then be used to impose or overcome Coulomb blockade in one of these SETs, switching the current into one or the other output branch. The 'branch switch', in contrast, uses two gate electrodes on each branch to create two tunnel barriers and form a single-island SET on each branch. Here, the gate voltages can be used to turn one or the other SET 'on' or 'off' and switch the current. Hasegawa and Kasai (Hasegawa and Kasai, 2001) proposed an implementation of these devices in III-V materials, using InGaAs or AlGaAs/GaAs/AlGaAs nanowires with Schottky in-plane or wrap gates, and in Si, using nanowires defined in SOI material.

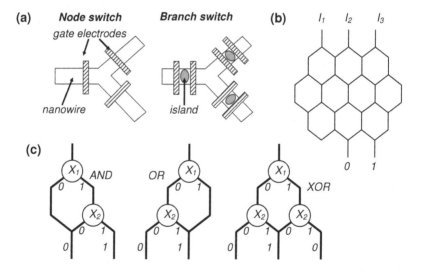

Fig. 6.16 BDD logic circuits using hexagonal arrays (after Hasegawa and Kasai, 2001). (a) Schematic diagram of switches. (b) Hexagonal array circuit tree. (c) Arrangement of AND, OR and XOR gates.

This basic Y-shaped two-way switch, combined with intermediate branches, can then be used to form a large hexagonal network of BDD nodes. Figure 6.16(b) shows a network with three input branches I_1, I_2 and I_3, a section of a hexagonal network, and the usual two output branches. Such a network directly interconnects the nodes, removing any problems with source-drain contact integration between separate devices. A specific BDD logic circuit can then be fabricated using a section of the network. The gate terminals associated with the nodes can be used to turn sections of the network 'on' or 'off' to implement the BDD function. Ungated sections of the network serve as interconnects. Figure 6.16(c) shows possible layouts for a hexagonal network implementation of BDD AND, OR and XOR gates.

The initial implementations of this proposal (Kasai *et al.*, 2000; Kasai and Hasegawa, 2002) used etched nanowires in AlGaAs/GaAs/AlGaAs material, defined by e-beam lithography. A schematic of the device is shown in Fig. 6.17(a). Electrons are confined in the potential well formed by the GaAs, and the current is controlled using Schottky wrap gates. A SET is fabricated using two gates deposited on the nanowire. Kasai and Hasegawa fabricated and characterized an OR gate using nanowires ~500 nm in width. Figure 6.17(b) shows a schematic diagram of the circuit. A single-gate 'node' switch was used to form node X1 and a 'branch' switch was used for node X2. The operation of the circuit was simplified by observing that a current on branch 1 precluded a current on branch 0 and vice versa. Therefore, only one output branch was enough to establish circuit operation, e.g. the presence or absence of a current at the '1' terminal could be used for logic determination.

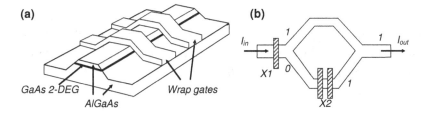

Fig. 6.17 BDD logic devices for hexagonal array circuits (after Hasegawa and Kasai, 2001). (a) Schematic diagram of a device. (b) Arrangement of devices within the circuit.

Kasai and Hasegawa operated the circuit from 1.6 to 120 K, with adjustments in the bias and gate voltages. At low temperature, single-electron operation was observed, and with increasing temperature, few-electron operation and finally, many-electron operation was observed. We note that this circuit does not use r.f. clocking signals, unlike the earlier work by Tsukagoshi et al. (Tsukagoshi et al., 1998). Precise control over the timing and electron packet size would require the addition of clocking signals to the circuit. A supply voltage V_{in} was necessary on the input branch to drive the current.

For operation at 1.6 K, the circuit used $V_{in} = 1$ mV. The gate voltage swings for nodes X1 and X2 were $\Delta V_{gX1} = 100$ mV and $\Delta V_{gX2} = 250$ mV. In contrast, for operation at 120 K, the circuit used $V_{in} = 0.2$ mV, $\Delta V_{gX1} = 1200$ mV and $\Delta V_{gX2} = 1000$ mV. For the SET switch, using an island capacitance of 70 aF, it was estimated that the switching speed $\tau = 1$ ns, and the power $P = 2 \times 10^{-13}$ W. This gave a power-delay product (PDP) = 2×10^{-22} J. For room temperature operation, smaller capacitances ~1 aF would be necessary. While this would reduce the PDP to ~10^{-20} J, the value would still be three orders of magnitude better than CMOS devices.

Implementation of the hexagonal network scheme in III-V materials raises the possibility of self-assembling parts of the circuit. This provides a means to fabricate SETs with better uniformity of the electrical characteristics, a necessary requirement for an integrated circuit. Nakajima et al. (Nakajima et al., 1999b, 2001) have used selective-area metal-organic vapour-phase epitaxy (SA-MOVPE) to create AlGaAs/GaAs/AlGaAs SETs with precisely grown device dimensions. These devices were grown using a SiON masking layer, defined on a (001) GaAs substrate, with the edges of the mask forming a zigzag pattern along the [100] and [110] directions. MOVPE growth at 700°C of an AlGaAs/GaAs/AlGaAs heterostructure on the masked substrate led to a nano-faceted structure which directly formed a diamond-shaped GaAs QD with an effective diameter ~60 nm, placed between source and drain leads.

The SA-MOVPE technique has been used to fabricate a Y-shaped two-way switch (Nakajima et al., 2002b), a hexagonal array AND/NAND gate (Nakajima et al., 2003) and a hexagonal array 1-bit adder (Miyoshi *et al.*, 2005). The more complex AND/NAND and adder circuits are constructed by using multiple elements of the hexagonal array. These circuits operated at 1.8 K, due to the rather large size of the QDs. However, with dots ~10 nm in size, it was expected that operation at 77 K may be possible.

6.3.2 Implementation of BDD logic circuits in silicon

Following the initial implementation of BDD single-electron logic circuits in δ-doped GaAs material (Tsukagoshi *et al.*, 1998), various implementations of these circuits in SOI material were demonstrated. Ono *et al.* (Ono *et al.*, 2000b) fabricated a two-way switching device, consisting of two SETs with a common source. The SETs were defined in SOI material using V-PADOX (See Chapter 3). The gate voltages of the SETs were used to switch the current from the input, common source terminal, to one of the output-drain terminals, at temperatures up to 40 K. Here, the source was grounded and a drain voltage applied to both SETs to drive the current through the circuit.

Radio-frequency clocking signals for timing and control over the electron packet size (Asahi *et al.*, 1997; Tsukagoshi *et al.*, 1998; Tsukagoshi and Nakazato, 1998) were not used in this circuit. The smaller island size possible in silicon SETs in comparison with GaAs SETs led to a higher temperature of operation. Ono and Takahashi (Ono and Takahashi, 2000) have also demonstrated the half-sum and carry-out of the half-adder, again using a two SET circuit. He *et al.* (He *et al.*, 2004a) have demonstrated an r.f.-clocked BDD two-way switch using silicon nanowire SETs, where few-electron packets are switched at 4.2 K. The operation of this device has been extended to demonstrate a universal three-way electron switch, with electron packets as small as two electrons transferred in any direction through the circuit (He *et al.*, 2004b). This device improves the performance of the BDD logic node almost to the ideal level of a single-electron messenger. The device may also be used to transfer an electron between different BDD logic circuits, e.g. between the input branches of BDD gates preliminary to the electron being injected into the logic circuit.

In more recent work, Saitoh et al. (Saitoh et al., 2005) have demonstrated a two-way current switch operating at room temperature using ultra-narrow silicon nanowire single-hole transistors (SHTs). This device uses two SHTs, each with a silicon channel less than 3 nm in width and height, and 200 nm long. Current oscillations with a very high peak-to-valley ratio >1000 were observed in the SHTs at 300 K. A single top-gate, coupled to both SHTs, could be used to switch the input current. A high on-off ratio of 8 was observed in the device at 300 K.

We now discuss the fabrication and electrical characterization of the two-way switch, and the universal three-way switch, of He et al. (He et al., 2004a; He et al., 2004b) in detail.

6.3.2.1 Two-way BDD switch using silicon nanowire SETs

He et al. (He et al., 2004a) have fabricated a two-way switch (the basic BDD node) in SOI material, based on silicon nanowire SET bi-directional electron pumps (Altebaeumer et al., 2001d; Altebaeumer and Ahmed, 2001a). Each nanowire SET uses two in-plane side-gates to control the current. The full two-way switch uses three SETs, connected to form two electron pumps in parallel. The device could be used to switch few-electron packets from the input or 'entry' branch into one of two output or 'exit' branches at 4.2 K.

Figure 6.18(a) shows the circuit diagram of the switch, with one 'entry branch' and two output branches, branch 1 and branch 0. The currents I_e, I_1 and I_0 flow in these branches respectively. The current polarities are taken to be positive when a current enters a branch. The SETs form two separate electron pumps. SET1 and SET2 form electron pump 1 and SET1 and SET3 form electron pump 2. The three SETs are connected to each other at a central node. The circuit is operated with all the branches grounded, and there is no need for a supply voltage. A sine wave r.f. signal, capacitively coupled to the central node, drives the circuit. In the standard manner of BDD logic, if the current flows from the entry branch to branch 1, then this represents logic 1 and if the current flows to branch 0, then this represents logic 0. The input signal X_1 is applied via the common gate voltage (V_{gcom}) to SET2 and SET3, and switches one or the other of these SETs 'on' or 'off'.

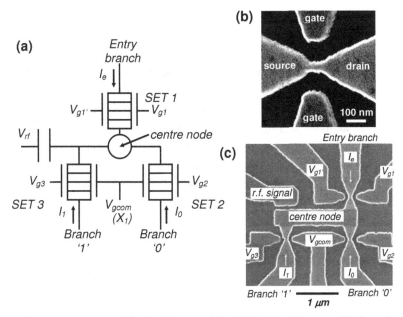

Fig. 6.18 Two-way switch in SOI material. (a) Circuit diagram. (b) Scanning electron micrograph of SET. (c) Scanning electron micrograph of the switch. (Reprinted from He *et al.*, *Microelectronic Engineering*. 'Two-way switch for binary decision diagram logic using silicon single-electron transistors'. Vol. 73–74, pp. 712–718. Copyright 2004, with permission from Elsevier).

The circuit was fabricated in SOI material, using a fabrication process similar to that described for silicon few-electron pumps in Chapter 5. The top silicon layer of the SOI material was 40 nm thick, and heavily doped n-type with phosphorus (doping concentration 2×10^{19} cm^{-3}). The underlying buried oxide layer was 400 nm thick. E-beam lithography in PMMA resist was used to define the SET patterns. A 30 nm-thick Al etch mask was then deposited using thermal evaporation and lift-off.

Reactive-ion etching in SiCl$_4$ plasma transferred the mask pattern into the top silicon layer. The Al etch mask was then removed by wet etching. The circuit was oxidized in dry O$_2$ for 45 minutes at 1000°C, to reduce nanowire cross-sections and passivate the etched silicon surfaces. Figures 6.18(b) and (c) show scanning electron micrographs (before oxidation) of a single SET, and of the full circuit, respectively. Before oxidation, the nanowires were ~50 nm wide and ~100 nm long. The width of the silicon region in the nanowire was reduced to ~20 nm by the oxidation. Finally, optical lithography was used to define the peripheral Cr/Ag contacts.

Figure 6.19 shows the I-V characteristics of one of the SETs (SET3) at 4.2 K. As there were always two SETs along a current path, the side-gate of one of the SETs was biased at a point where the conductance was large, e.g. at a large single-electron conductance peak. The SET then simply acted as a resistor and it was possible to characterize the other SET. The applied nanowire voltage included a part dropped across the 'resistor' SET. The characteristics were measured with a voltage V_{ds} applied between the entry branch and branch 1, and a voltage V_{gcom} applied to the common gate.

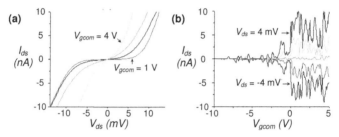

Fig. 6.19 Electrical characteristics of SET3 (see Fig. 6.18). (a) I_{ds} vs. V_{ds}. (b) Single-electron oscillations, I_{ds} vs. V_{gcom}. (Reprinted from He et al., Microelectronic Engineering. 'Two-way switch for binary decision diagram logic using silicon single-electron transistors'. Vol. 73–74, pp. 712–718. Copyright 2004, with permission from Elsevier).

Throughout the measurement, the other side-gate voltage of SET3, $V_{g3} = 0$ V. SET1 was biased at a wide single-electron conductance peak, at $V_{g1} = 0$ V and $V_{g1'} = -10$ V. Figure 6.19(a) shows the current in SET3, I_{ds} vs. V_{ds}, as V_{gcom} is varied from 1 to 4 V. The maximum width of the Coulomb gap is ~8 mV, at $V_{gcom} = 1$ V. Figure 6.19(b) shows the single-electron conductance oscillations in I_{ds} vs. V_{gcom}, as V_{ds} is varied from -4 mV to 4 mV in 2 mV steps. A period ΔV_{gcom} ~0.6 V can be identified, although this is complicated by additional oscillations in the MTJ formed by the nanowire. ΔV_{gcom} can be associated with an island with a capacitance $C_g = e/\Delta V_{gcom} = 0.27$ aF.

SET1 and SET2, or SET1 and SET3, each form a bi-directional few-electron pump. The operation of these devices was discussed in detail in Chapter 5. The pump is driven by a sine wave r.f. signal capacitively coupled to the central node. This alters the potential of the central node in a manner such that the Coulomb blockade in the first of the SETs in a pair (SET1) is overcome only in the first half of the r.f. cycle, and the Coulomb blockade in the other SET (SET2, in the pair SET1 and SET2) is overcome only in the second half of the r.f. cycle. Consequently, the SETs successively turn 'on' and 'off', transferring a small packet of electrons across the pump in each r.f. cycle. By adjusting the SET gate voltages, the position of the operating point of each SET can be adjusted and the polarity of the current switched, i.e. electrons can be transferred in either direction across the pump. The electron packet size depends on the amplitude of the r.f. signal, and on the charging and discharging 'RC' time of the central node across each SET.

Figure 6.20 shows the characteristics of the two-way switch at 4.2 K. The entry branch, branch 0 and branch 1 are all grounded and a 3 MHz and 50 mV peak-to-peak sine wave r.f. signal is applied to drive the circuit. Figure 6.20(a–c) shows the currents I_e in the entry branch, I_1 in branch 1, and I_0 in branch 0, respectively, as a function of V_{gcom} and at $V_{g1} = -5.6$ V. Electron pump oscillations are observed in each of the graphs. Both positive and negative current values are observed, demonstrating the bi-directionality of electron transfer in the circuit.

We note that the sum of I_1 and I_0 is equal to $-I_e$, as expected from Fig. 6.18(a). We now identify points where the current is switched into one or the other branch. At $V_{gcom} = 8$ V, $I_1 \approx -0.3$ nA, $I_e \approx 0.3$ nA and $I_0 \approx 0$ nA. This implies that the current flows from the entry branch into branch 1. Using the equation for the electron pump current $I = nef$, where n is the number of electrons and f is the pumping frequency, ~600 electrons are transferred per r.f. cycle at this biasing point. The condition of SET3 and SET2 is marked in Fig. 6.20(b–c). The Coulomb gap is reduced in SET3 and a wide Coulomb gap exists in SET2, implying that SET3 is 'on' and SET2 is 'off'. Conversely, at $V_{gcom} = 5$ V, the current flows from the entry branch to branch 0. Using these two values of V_{gcom} for an input switching signal, i.e. using $V_{gcom} = 8$ V to represent $X_1 = 1$, and using

$V_{gcom} = 5$ V to represent $X_1 = 0$, the circuit operates as a two-way switch. Other points may also be chosen for switching operations.

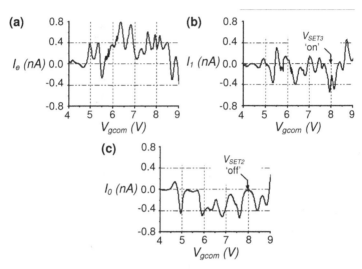

Fig. 6.20 Electrical characteristics of the two-way switch (see Fig. 6.18). (a) Current I_e in entry branch. (b) Current I_1 in branch 1. (c) Current I_0 in branch 0. (Reprinted from He et al., Microelectronic Engineering. 'Two-way switch for binary decision diagram logic using silicon single-electron transistors'. Vol. 73–74, pp. 712–718. Copyright 2004, with permission from Elsevier).

6.3.2.2 Extension to a 'universal' three-way switch

A two-way switch similar to the device discussed in the preceding section (He et al., 2004a) can be operated in a more generalized manner as a 'universal' three-way switch (Fig. 6.21) (He et al., 2004b), where few-electron packets can be switched in *any* direction between the three branches connected to the central node. This device also uses bi-directional electron pumps formed by three silicon nanowire SETs and an r.f. signal to transfer the few-electrons packets. However, better gate control over all three SETs is necessary. The switching of packets ~10 electrons in size, and for specific biasing points as small as two electrons in size, has been demonstrated in this device. Device operation is then close to the 'ideal' packet size of only one electron. The device forms an improved BDD logic node and may also be used to transfer single electrons between different BDD logic circuits.

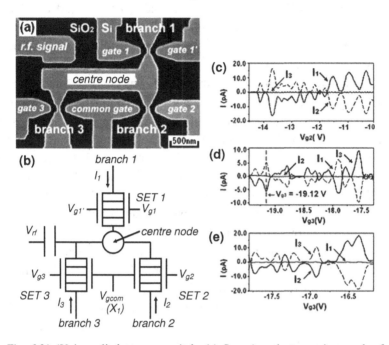

Fig. 6.21 'Universal' three-way switch. (a) Scanning electron micrograph of device. (b) Circuit diagram. (c) Current switching between branch 1 and 2. (d) Current switching between branch 1 and 3. (e) Current switching between branch 2 and 3. (Reprinted with permission from He *et al.*, *Appl. Phys. Lett.* **85**, 308 [2004]. Copyright 2004, American Institute of Physics).

The device layout is similar to the two-way switch of Fig. 6.18(b), and the same fabrication process is used to define the device. Figure 6.21(a) shows a SEM image of the device. Three silicon nanowire SETs, connected to a central node, form three branches of the switch, branch 1, branch 2 and branch 3 (Fig. 6.21[b]). While the basic circuit is similar to the two-way switch of Fig. 6.18(a), the circuit is operated differently, in a more generalized manner. Currents I_1, I_2 and I_3 flow in the three branches, respectively, taken to have positive polarity when currents enter a branch. Unlike the two-way switch, in the universal switch, *any* series combination of two SETs can be operated as a bi-directional electron pump. Therefore, there are three electron pumps, pump 'A'

(SET1 and SET2), pump 'B' (SET1 and SET3) and pump 'C' (SET2 and SET3). In practice, this requires that all the SETs have broadly similar characteristics. All three branches are connected to ground and an r.f. signal, capacitively coupled to the centre node, drives the pumps.

Using the control gate voltages V_{g1}, V_{g2} and V_{g3} to adjust the Coulomb gap in the SETs and select specific electron pumps, few-electron packets can be transferred to any of the three branches, in any direction. The remaining gates ($V_{g1'}$ and V_{gcom}) are not essential for device operation. Figures 6.21(c–e) show the transfer of electrons through the switch using a 3 MHz and 200 mV peak-to-peak r.f. signal, along the routes formed by branch 1 and 2, branch 1 and 3, and branch 2 and 3, respectively. In Fig. 6.21(c), I_1, I_2 and I_3 are plotted as V_{g2} is varied from -10 V to -14.5 V ($V_{g1} = -17$ V and $V_{g3} = -18.9$ V are constant). We observe that $I_1 \approx -I_2$, and $I_3 \approx 0$ pA. When I_1 is positive, packets of electrons (negative charge) are transferred from branch 2 to branch 1. Conversely, when I_1 is negative, electrons are transferred from branch 1 to branch 2. Similarly, Fig. 6.21(d) shows electron transfer along branch 1 and branch 3 and Fig. 6.21(e) shows electron transfer along branch 2 and branch 3.

In Fig. 6.21(d) at $V_{g3} = -19.1$ V, the current values are $I_1 = 5.00$ pA, $I_2 = 0.13$ pA and $I_3 = -5.60$ pA. Here, $I_1 + I_2 + I_3 = -0.47$ pA, within the noise level. From the electron pump equation $I = nef$, a current of ~5 pA corresponds to only ~10 electrons pumped per r.f. cycle. The smallest current peaks in the graphs are only ~1 pA in magnitude, corresponding to only two electrons pumped per r.f. cycle. For these peaks, device operation tends towards the ideal of one electron pumped per cycle.

6.4 Quantum Cellular Automaton Circuits

QCA may provide a very different approach to logic circuits using single-electron or QD devices (Lent *et al.*, 1993a, 1993b; Tougaw *et al.*, 1993, 1994, 1996). A QCA consists of an array of identical cells, each consisting of a specific arrangement of QDs or single-electron charging islands. The electrostatic coupling between the QDs forces a cell into one of two possible charge configurations, with two different polarizations. These two charge configurations are used to represent the '1' and '0' states. Data is input into the array by setting the state of cells along one of the edges of the array. Depending on this edge configuration, the various cells forming the QCA array switch each other into the '1' or '0' states, such that the entire array relaxes into the lowest energy configuration, i.e. the ground state. The array is designed such that with the array in the ground state, the states of cells along a different edge give the output data. The process is also referred to as 'computing with the ground state' (Lent *et al.*, 1993b). At present, most of the work on QCAs is theoretical, and there are only a few demonstrations of QCA cells, discussed later in this section.

The basic QCA cell, referred to as the 'standard cell', consists of five electrostatically interacting QDs, arranged as shown in Fig. 6.22(a) (Lent *et al.*, 1993b). The QDs are charged with two extra electrons, which can tunnel between the QDs. The four QDs at the corners can interact with each other via the fifth QD at the centre of the cell, and directly along the edges of the cell. The former (nearest neighbour) interaction is more significant and the direct (next nearest neighbour) interaction can often be neglected as an approximation (Tougaw *et al.*, 1993). The interaction between the QDs leads to tunnelling of the two electrons such that they occupy one of two possible stable charge configurations, along the cell diagonals (Fig. 6.22[a]). These configurations can be used to represent the '1' and '0' states. The two charge configurations are associated with two possible polarizations of the cell.

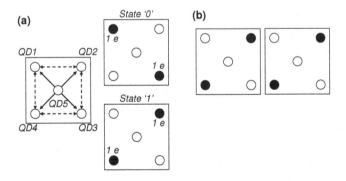

Fig. 6.22 (a) Basic QCA cell, and '0' and '1' states of the cell. (b) Interaction between cells.

For an isolated cell, both configurations are equally likely and are degenerate in energy. However, if two cells are placed near to each other such that they interact electrostatically (Fig. 6.22[b]), then the polarization of one of the cells affects the other so that the degeneracy is removed, and a lower energy ground state is formed for similar polarizations in both cells. Tougaw *et al.* (Tougaw *et al.*, 1993) have theoretically demonstrated that for the five-QD cell, this switching effect is very strong. Simpler configurations of cells are also possible, e.g. where the central QD is removed, only two tunnel-coupled QDs are used, or elongated QDs forming 'quantum dashes' (Bakshi *et al.*, 1991) are used.

Tougaw *et al.* find that, while these configurations simplify the design of the cell and reduce the number of QDs per cell, the cell–cell interaction is greatly weakened. The four-cell configuration has been investigated by Wu *et al.* (Wu *et al.*, 1997) with a view to implementation by SETs. Alternative forms of cells may also be possible, e.g. with single electrons stored in cells formed by one QD each, the spin of the electron may be used to switch the cells (Bandyopadhyay *et al.*, 1994).

It is possible to define various components of the QCA array, e.g. wires, NOT, AND and OR gates, using arrangements of individual cells (Tougaw *et al.*, 1994). The QCA wire (Fig. 6.23[a]) simply consists of a line of cells, where the input polarization switches the polarization of the next cell, the process continuing until all cells are switched. Fan-out of a single line into two is easily arranged by a T-shaped configuration. If two cells are arranged diagonally (Fig. 6.23[b]), the polarization of the first switches the other into the opposite polarization, forming an inverter. Programmable two input OR and AND gates are shown in Fig. 6.23(c–d). The basic configuration of both gates is similar, forming a cross where one of the arms is used as a programme line, two arms form the input lines, and the fourth forms the output line. For the OR gate, the programme line is set to '1'. The polarization of the central cell then depends on the net polarization of the two input arms and the programme line, i.e. a 'majority logic' decision. Therefore, for inputs (1, 1), (1, 0) or (0, 1), in combination with the '1' on the programme line, the net polarization effect on the central cell is '1' and the cell is switched to '1', leading to an output of '1'.

For input (0, 0), the net polarization effect on the central cell is a '0', leading to an output of '0'. This forms an OR gate. In contrast, if the programme line is set to '0', the configuration forms an AND gate (Fig. 6.23[d]). Once a QCA has been designed to implement a given logic function, the input data sets the boundary conditions at one edge (Fig. 6.23[e]). Depending on this data, the QCA is expected to relax into its ground state. The output data can then be read from a different QCA edge. Configurations for other logic functions, e.g. for a XOR gate and for a 1-bit full adder, have also been suggested (Tougaw *et al.*, 1994).

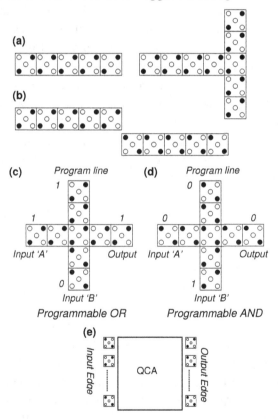

Fig. 6.23 QCA configurations. (a) QCA Wire. (b) QCA Inverter. (c) Programmable OR gate. (d) Programmable AND gate. (e) Data input and output to a QCA.

The implementation of QCA logic functions would require very large numbers of cells, e.g. the 1-bit full adder of Tougaw et al. (Tougaw et al., 1994) requires 192 cells. Each cell requires at least one QD (five QDs in the 'standard' cell), and control over the position of these QDs. Both of these issues form major challenges, difficult to overcome at present. The QCA cells also have no inherent gain, and this leads to poor isolation of the output from the input. Furthermore, the QCA must be able to relax into the ground state, and the process must not stop at an intermediate state. The operating temperature of the QCA depends on the operating temperature of the QDs, and room-temperature operation would require sub-10 nm QDs.

However, if these challenges can be overcome, a QCA approach may have various advantages. A QCA does not require physical interconnects between cells because the cells switch through electrostatic interaction. Data input is at the edge of the array and interconnects to cells within the array are unnecessary. The QCA uses closely packed QDs and can have a very high packing density. Power dissipation is greatly reduced as this is only associated with the tunnelling of the two electrons within each cell. The switching speed depends on the relaxation time of the cells, of the order of picoseconds. Each cell also has two stable configurations, inherently forming a memory cell.

A QCA cell has been fabricated by Orlov et al. (Orlov et al., 1997) using four Al islands, connected by AlO_x tunnel junctions of area ~60 × 60 nm. The circuit operated at a low temperature, below 50 mK, due to the comparatively large island size. Figure 6.24(a) shows the circuit diagram of the cell. The islands D1, D2, D3 and D4 form a four-site QCA cell, with the two polarization states shown in Fig. 6.24(b). Islands D1 and D2 form two SETs, with the central terminal connected to ground. The gate voltages V_1 and V_2 can be used to adjust the number of electrons (m1, m2) on these two islands. The islands D3 and D4 are tunnel-coupled to each other, and form a double dot. The number of electrons (n1, n2) on the dots D1 and D2 can be controlled using the gate voltages V_3 and V_4. As a function of these voltages, the electron number is stable within a set of hexagonal regions, schematically shown in Fig. 6.24(c). Either n1 or n2 changes by one if the gate voltages are adjusted such that the operating point of the double dot moves from a hexagonal region to a neighbouring region, e.g. if V_3 and V_4 are adjusted such that the double dot moves along the arrowed line from the (1, 0) stability region to the (0, 1) region, an electron is exchanged between islands D3 and D4.

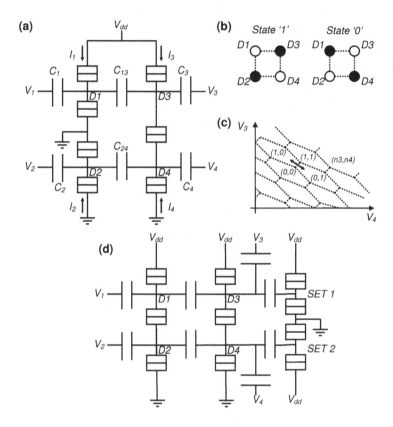

Fig. 6.24 Implementation of basic QCA cell (after Orlov et al., 1997). (a) Circuit diagram of cell. (b) Polarization states. (c) Stability diagram. (d) Circuit diagram of QCA cell with electrometers for sensing (after Amlani et al., 1998).

Orlov et al. (Orlov et al., 1997) demonstrated that their circuit behaved as a four-site QCA cell as follows. V_3 and V_4 were set such that for zero extra electrons on D1 and D2, i.e. (m1, m2) = (0, 0), D3 and D4 were biased at the boundary between the (1, 0) and (0, 1) regions, at point 'A'. Using V_1 and V_2, one electron was then added to the island D1 such that the electron number (m1, m2) was (1, 0). This electron biased

D3 and D4, shifting the stability regions shown in Fig. 6.24(c) along the arrowed line such that an electron was transferred from D3 to D4, i.e. the configuration (n1, n2) = (0, 1). Similarly, a configuration (m1, m2) = (0, 1) for D1 and D2 led to the transfer of an electron from D4 to D3, i.e. the configuration (n1, n2) = (1, 0). This demonstrated the formation of the two polarization states required for a QCA cell (Fig. 6.24[b]).

In the QCA cell of Orlov *et al.*, the detection of the charge state of D3 and D4 required the measurement of the corresponding stability diagram by varying V_3 and V_4, and the application of cancellation voltages at V_1 and V_2 to prevent a change in the charge configuration of D1 and D2. Amlani *et al.* (Amlani *et al.*, 1998) extended the design of the cell to include two SET electrometers, capacitively coupled to D3 and D4, respectively (Fig. 6.24[d]). These electrometers could sense the switching of the charge states of D3 and D4, without the need to characterize the stability diagram of D3 and D4 for the verification of the charge states. In addition, a tunnel junction was used to connect D1 and D2 in a manner similar to D3 and D4, creating a symmetrical cell design.

6.5 Single-Electron Parametron

The QCA cells discussed above are not the only means to use a polarized cell to encode logic bits. Korotkov and Likharev (Korotkov and Likharev, 1998) have proposed a 'single-electron parametron' cell, where three islands are used to create two different polarizations. A schematic diagram of the cell is shown in Fig. 6.25(a). The cell consists of three islands, tunnel-coupled to their nearest neighbour. The central island is spatially shifted slightly off the line joining the centres of the other two islands. In Fig. 6.25(a), this is in the positive *y* direction. An electric field F_C is applied along the *y*-axis, lowering the Fermi energy of the central island. In addition, a small electric field F_S is applied such that the Fermi energy of the right island is lower than that of the left (Fig. 6.25[b]). Now, if one electron is added to the cell, due to the effect of F_C, it charges the central island. If we then reduce the magnitude of F_C, the Fermi level of the central island moves up in energy until the electron tunnels into the right island. However, reversing the direction of F_S

would cause the electron to tunnel into the left island. Reducing F_C to zero traps the electron in one of the outside islands. This creates two different polarizations of the cell, forming the '1' and '0' states (Fig. 6.25[c]).

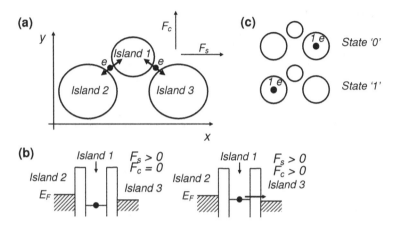

Fig. 6.25 Single-electron parametron (after Korotkov and Likharev, 1998). (a) Schematic diagram of the cell. (b) Effect of electric field on energy band diagram (c) Cell states '0' and '1'.

The two polarizations of the cell create a small field F_S, along either the negative or positive x direction. This field can then be used to control the switching of other parametron cells. An array of parametron cells can be used to implement logic functions, e.g. arrays for a shift register, a NAND gate and a NOR gate have been developed (Korotkov and Likharev, 1998; Likharev, 1999). The parametron approach has the advantage of a wireless approach, with the circuit driven by the clocking electric field F_C. The power dissipation in an array can theoretically be very low, less than $(\ln 2)k_B T$ per bit. This raises the possibility of extremely high levels of circuit integration. However, Kortokov and Likharev estimate that, even with 5 nm-diameter islands, the maximum operating temperature is only ~15 K. Room temperature operation may require even smaller, sub-nanometre scale islands.

The fabrication of a parametron array also requires control over the size and position of nanoscale islands – a difficult task. However, a single parametron cell has been fabricated using islands defined in the top silicon layer of SOI material by e-beam lithography (Emiroglu *et al.*, 2002, 2003). Here, the right and left islands are connected by a silicon nanowire, and the central island lies within the MTJ formed by the nanowire. The polarization of the cell at 4.2 K can be switched by capacitively coupled gate electrodes, and sensed using a SET electrometer.

Bibliography

Akers S.B. (1978). Binary decision diagrams. *IEEE Trans Computers* C-27: 509-515.
Ali D. and Ahmed H. (1994). Coulomb blockade in a silicon tunnel junction device 64: 2119-2120.
Alt'shuler B.L. (1985). Fluctuations in the extrinsic conductivity of disordered conductors. *JETP Lett* 41: 648-651.
Altebaeumer T. and Ahmed H. (2001a). Characteristics of electron pump circuits using silicon multiple tunnel junctions. *Jpn J Appl Phys* 40: 80-82.
Altebaeumer T. and Ahmed H. (2001b). Silicon nanowires and their application in bi-directional electron pumps. *Microelectronic Engineering* 57-58: 1029-1033.
Altebaeumer T. and Ahmed H. (2001c). Electrical characteristics of two Coulomb blockade devices driven by a periodically oscillating potential. *J Appl Phys* 90: 1350-1356.
Altebaeumer T. and Ahmed H. (2002). Performance of silicon based bi-directional electron pumps consisting of two Coulomb blockade devices. *Jpn J Appl Phys* 41: 2694-2697.
Altebaeumer T. and Ahmed H. (2003). Tunnel barrier formation in silicon nanowires. *Jpn J Appl Phys* 42: 414-417.
Altebaeumer T., Amakawa S. and Ahmed H. (2001d). Characteristics of two Coulomb blockade transistors separated by an island to which an oscillating potential is applied: Theory and experiment. *Appl Phys Lett* 79: 533-535.
Altebaeumer T., Amakawa S. and Ahmed H. (2003). Cross-coupling in Coulomb blockade circuits: Bidirectional electron pump. *J Appl Phys* 94: 3194-3200.
Amakawa S., Majima H., Fukui H., Fujishima M. and Hoh K. (1998). Single-electron circuit simulation. *IEICE Trans Electron* E81-C: 21-29.
Amakawa S., Mizuta H. and Nakazato K. (2001). Analysis of multiphase clocked electron pumps consisting of single-electron transistors. *J Appl Phys* 89: 5001-5008.
Amlani I., Orlov A.O., Snider G.L., Lent C.S. and Bernstein, G.H. (1998). Demonstration of a six-dot quantum cellular automata system. *Appl Phys Lett* 72: 2179-2181.
Amman M., Ben-Jacob E. and Mullen K. (1989). Charge solitons in 1-D array of mesoscopic tunnel junctions. *Phys. Lett. A* , 142: 431-437.

Ancona M.G. (1996). Design of computationally useful single-electron digital circuits. *J Appl Phys* 79: 526–539.
Ancona M. G. and Rendell R. W. (1995). Simple computation using Coulomb blockade-based tunneling arrays. *J Appl Phys* 77: 393–395.
Asahi N., Akazawa M. and Amemiya Y. (1995). Binary-decision-diagram device. *IEEE Trans Electron Devices* 42: 1999–2003.
Asahi N., Akazawa M. and Amemiya Y. (1997). Single-electron logic device based on the binary decision diagram. *IEEE Trans Electron Devices* 44: 1109–1117.
Asahi N., Akazawa M. and Amemiya Y. (1998). Single-electron logic systems based on the binary decision diagram. *IEICE Trans Electron* E81-C: 49–56.
Ashori R.C. (1996). Electrons in artificial atoms. *Nature* 379: 413–419.
Augke R., Eberhardt W., Single C., Prins F.E., Wharam D.A. and Kern, D.P. (2000). Doped silicon single electron transistors with single island characteristics. *Appl Phys Lett* 76: 2065–2067.
Averin D.V. and Likharev K.K. (1986). Coulomb blockade of single-electron tunneling, and coherent oscillations in small tunnel junctions. *J Low Temp Phys* 62: 345–373.
Averin D.V. and Likharev K.K. (1991). Single electronics: A correlated transfer of single electrons and Cooper pairs in systems of small tunnel junctions in Altshuler B., Lee P. and Webb R. (eds) *Mesoscopic Phenomena in Solids* Elsevier, Amsterdam.
Averin D.V. and Likharev K.K. (1992). Possible applications of the single charge tunneling in Grabert H. and Devoret M.H. (eds) *Single Charge Tunneling* Plenum, New York.
Averin D.V. and Nazarov Y.V. (1992). Macroscopic quantum tunnelling of charge and co-tunnelling in Grabert H. and Devoret M.H. (eds) *Single Charge Tunneling* Plenum, New York.
Averin D.V. and Odintsov A.A. (1989). Macroscopic quantum tunneling of the electric charge in small tunnel junctions. *Phys Lett A* 140: 251–257.
Averin D.V., Korotkov A.N. and Likharev K.K. (1991). Theory of single-electron charging of quantum wells and dots. *Phys Rev B* 44: 6199–6211.
Averin D.V., Odintsov A.A. and Vyshenskii S.V. (1993). Ultimate accuracy of single-electron dc current standards. *J Appl Phys* 73: 1297–1308.
Baccarani G., Ricco B. and Spandini, G. J. (1978). Transport properties of polycrystalline silicon films. *J Appl Phys* 49: 5565–5570.
Bakhvalov N.S., Kazacha G.S., Likharev K.K. and Serdyukova, S.I. (1989). Single-electron solitons in one-dimensional tunnel structures. *Sov Phys JETP* 68: 581–587.
Bakshi P., Broido D.A. and Kempa K. (1991). Spontaneous polarization of electrons in quantum dashes. *J Appl Phys* 70: 5150–5152.
Bandyopadhyay S., Das B. and Miller A.E. (1994). Supercomputing with spin-polarized single electrons in a quantum coupled architecture. *Nanotechnology* 5: 113–133.
Barner J.B. and Ruggiero S.T. (1987). Observation of the incremental charging of Ag particles by single electrons. *Phys Rev Lett* 59: 807–810.

Beenakker C.W.J. (1991). Theory of Coulomb-blockade oscillations in the conductance of a quantum dot. *Phys Rev B* 44: 1646–1656.
Ben-Jacob E. and Gefen Y. (1985). New quantum oscillations in current driven small junctions. *Phys Lett A* 108: 289–292.
Ben-Jacob E., Mullen K. and Amman M. (1989). Charge effect solitons. *Phys Lett A* 135: 390–396.
Blick R.H., Pfannkuche D., Haug R. J., Klitzing K.v. and Eberl K. (1998). Formation of a coherent mode in a double quantum dot. *Phys Rev Lett* 80: 4032–4035.
Brus L.E., Szajowski P.F., Wilson W.L., Haris T.D., Shuppler S. and Citrin, P.H. (1995). Electronic spectroscopy and photophysics of Si nanocrystals: Relationship to bulk c-Si and porous Si. *J Am Chem Soc* 117: 2915–2922.
Bryant R.E. (1986). Graph-based algorithms for boolean function manipulation. *IEEE Trans Computers* C-35: 677–691.
Cain P.A., Ahmed H. and Williams D.A. (2001). Conductance peak splitting in hole transport through a SiGe double quantum dot. *Appl Phys Lett* 78: 3624–3626.
Cain P.A., Ahmed H., Williams D.A. and Bonar J.M. (2000). Hole transport through single and double SiGe quantum dots. *Appl Phys Lett* 77: 3415–3417.
Chen C.D., Nakamura Y. and Tsai J.S. (1997). Aluminum single-electron nonvolatile floating gate memory cell. *Appl Phys Lett* 71: 2038–2040.
Chen R.H. and Likharev K.K. (1998). Multiple-junction single-electron transistors for digital applications. *Appl Phys Lett* 72: 61–63.
Chen R.H., Korotkov A.N. and Likharev K.K. (1996). Single-electron transistor logic. *Appl Phys Lett* 68: 1954–1956.
Choi B.H., Hwang S.W., Kim I.G., Shin H.C., Kim Y. and Kim E.K. (1998). Fabrication and room-temperature characterization of a silicon self-assembled quantum-dot transistor. *Appl Phys Lett* 73: 3129–3131.
Cleland A.N., Schmidt J.M. and Clarke J. (1990). Charge fluctuations in small-capacitance junctions. *Phys Rev Lett* 64: 1565–1568.
Colinge, J. P. (1997). *Silicon-on-insulator technology: Materials to VLSI*, Kluwer Academic Publishers, Boston.
Colli, A., Fasoli, A., Beecher, P., Servati, P., Pisana, S., Fu, Y., Flewitt, A. J., Milne, W. I., Robertson, J., Ducati, C., De Franceschi, S., Hofmann, S. and Ferrari, A. C. (2007). Thermal and chemical vapor deposition of Si nanowires: Shape control, dispersion, and electrical properties, *J Appl Phys* 102: 34302–34314.
Cower M.E. and Sedgwick T.O. (1972). Chemical vapor deposited polycrystalline silicon. *J Electrochem Soc* 119: 1565–1570.
Cui Y., Lauhon L.J., Gudiksen M.S., Wang J. and Lieber C.M. (2001). Diameter-controlled synthesis of single-crystal silicon nanowires. *Appl Phys Lett* 78: 2214–2216.
Cui Y., Zhong Z., Wang D., Wang W.U. and Lieber C.M. (2003). High performance silicon nanowire field effect transistors. *Nano Lett* 3: 149–152.

Delsing P., Claeson T., Likharev K.K. and Kuzmin L.S. (1990). Observation of single-electron-tunneling oscillations. *Phys Rev B* 42: 7439–7449.

Delsing, P. (1992). One-dimensional arrays of small tunnel junctions, in *Single Charge Tunneling*, Eds. Grabert, H. and Devoret, M. H., Plenum, New York.

Delsing P., Likharev K.K., Kuzmin L.S. and Claeson T. (1989a). Effect of high-frequency electrodynamic environment on the single-electron tunneling in ultrasmall junctions. *Phys Rev Lett* 63: 1180–1183.

Delsing P., Likharev K.K., Kuzmin L.S. and Claeson T. (1989b). Time-correlated single-electron tunneling in one-dimensional arrays of ultrasmall tunnel junctions. *Phys Rev Lett* 63: 1861–1864.

Devoret M.H. and Grabert H. (1992). Introduction to single charge tunneling in Grabert H. and Devoret M.H. (eds) *Single Charge Tunneling* Plenum, New York.

Devoret M.H., Esteve D. and Urbina C. (1992). Single-electron transfer in metallic nanostructures. *Nature* 360: 547–553.

Devoret M.H., Esteve D., Grabert H., Ingold G.-L., Pothier H. and Urbina C. (1990). Effect of the electromagnetic environment on the Coulomb blockade in ultrasmall tunnel junctions. *Phys Rev Lett* 64: 1824–1827.

Dong Y., Wang C., Wehrenberg B.L. and Guyot-Sionnest P. (2004). Variable range hopping conduction in semiconductor nanocrystal solids. *Phys Rev Lett* 92: 216802–216805.

Dresselhaus P.D., Ji L., Han S., Lukens J.E. and Likharev K.K. (1994). Measurement of single electron lifetimes in a multijunction trap, *Phys Rev Lett* 72: 3226–3229.

Durrani, Z. A. K. (2003). Coulomb blockade, single-electron transistors and circuits in silicon, *Physica E* 17: 572–578.

Durrani, Z. A. K., Irvine, A. C. and Ahmed, H. (2000). Coulomb blockade memory using integrated single-electron transistor/metal–oxide-semiconductor transistor gain cells. *IEEE Trans Electron Devices* 47: 2334–2339.

Durrani, Z. A. K., Irvine, A. C., Ahmed, H. and Nakazato, K. (1999). A memory cell with single-electron and metal-oxide-semiconductor transistor integration, *Appl Phys Lett* 74: 1293–1295.

Durrani, Z. A. K., Kamiya, T., Tan, Y. T. and Ahmed, H. (2002). Single-electron charging in nanocrystalline silicon point-contacts, *Microelectronic Engineering* 63: 267–275.

Dutta A., Kimura M., Honda Y., Otobe M., Itoh A. and Oda S. (1997). Fabrication and electrical characteristics of single electron tunneling devices based on Si quantum dots prepared by plasma processing. *Jpn J Appl Phys* 36: 4038–4041.

Dutta A., Lee S.P., Hatatani S. and Oda S. (1999). Silicon-based single-electron memory using a multiple-tunnel junction fabricated by electron-beam direct writing. *Appl Phys Lett* 75: 1422–1424.

Dutta A., Lee S.P., Hayafune Y., Hatatani S. and Oda S. (2000). Single-electron tunneling devices based on silicon quantum dots fabricated by plasma process. *Jpn J Appl Phys* 39: 264–267.

Dutta A., Oda S., Fu Y. and Willander M. (2000). Electron transport in nanocrystalline Si based single electron transistors. *Jpn J Appl Phys* 39: 4647–4650.
Emiroglu E.G., Durrani Z.A.K., Hasko D.G. and Williams D.A. (2002). Single-electron parametron memory cell. *J Vac Sci Technol B* 20: 2806–2809.
Emiroglu E.G., Durrani Z.A.K., Hasko D.G. and Williams D.A. (2003). Silicon single-electron parametron cell for solid-state quantum information processing. *Microelectronic Engineering* 67–68: 755–762.
Evans G.J., Mizuta H. and Ahmed H. (2001). Modelling of structural and threshold voltage characteristics of randomly doped silicon nanowires in the Coulomb-blockade regime. *Jpn J Appl Phys* 40: 5837–5840.
Ferry D.K. and Goodnick S.M. (1997). *Transport in Nanostructures*. Cambridge University Press, Cambridge.
Fischetti M.V. and Laux S.E. (1988). Monte carlo analysis of electron transport in small semiconductor devices including band-structure and space-charge effects. *Phys Rev B* 38: 9721–9745.
Fonseca L.R.C., Korotkov A.N. and Likharev K.K. (1996a). A numerical study of the accuracy of single-electron current standards. *J Appl Phys* 79: 9155–9165.
Fonseca L.R.C., Korotkov A.N. and Likharev K.K. (1996b). Accuracy of the single-electron pump using an optimized step-like rf drive waveform. *Appl Phys Lett* 69: 1858–1860.
Fripp A.L. (1975). Dependence of resistivity on the doping level of polycrystalline silicon. *J Appl Phys* 46: 1240–1244.
Fujiwara A. and Takahashi Y. (2001). Manipulation of elementary charge in a silicon charge-coupled device. *Nature* 410: 560–562.
Fujiwara A., Takahashi Y., Murase K. and Tabe M. (1995). Time-resolved measurement of single-electron tunneling in a Si single-electron transistor with satellite Si islands. *Appl Phys Lett* 67: 2957–2959.
Fujiwara A., Takahashi Y., Namatsu H., Kurihara K. and Murase K. (1998). Suppression of effects of parasitic metal-oxide-semiconductor field-effect transistors on Si single-electron transistors. *Jpn J Appl Phys* 37: 3257–3263.
Fujiwara A., Zimmerman N.M., Ono Y. and Takahashi Y. (2004). Current quantization due to single-electron transfer in Si-wire charge-coupled devices. *Appl Phys Lett* 84: 1323–1325.
Fulton T.A. and Dolan G.J. (1987). Observation of single-electron charging effects in small tunnel junctions. *Phys Rev Lett* 59: 109–112.
Fulton T.A., Gammel P.L. and Dunkelberger L.N. (1991). Determination of Coulomb-blockade resistances and observation of the tunneling of single electrons in small-tunnel-junction circuits. *Phys Rev Lett* 67: 3148–3151.
Furuta Y., Mizuta H., Nakazato K., Kamiya T., Tan Y.T., Durrani Z.A.K. and Taniguchi K. (2002). Characterization of tunnel barriers in polycrystalline silicon point-contact single-electron transistors. *Jpn J Appl Phys* 41: 2675–2678.

Furuta Y., Mizuta H., Nakazato K., Tan Y.T., Kamiya T., Durrani Z.A.K., Ahmed H. and Taniguchi K. (2001). Carrier transport across a few grain boundaries in highly doped polycrystalline silicon. *Jpn J Appl Phys* 40, L615–L617 (2001).

Geerligs L.J., Anderegg V.F., Holweg P.A.M., Mooij J.E., Pothier H., Esteve D., Urbina C. and Devoret M.H. (1990). Frequency-locked turnstile device for single electrons. *Phys Rev Lett* 64: 2691–2694.

Geerligs L.J., Averin D.V. and Mooij J.E. (1990). Observation of macroscopic quantum tunneling through the Coulomb energy barrier. *Phys Rev Lett* 65: 3037–3040.

Geigenmüller U. and Schön G. (1989). Single-electron effects in arrays of normal tunnel junctions. *Europhys Lett* 10: 765–770.

Giaever I. and Zeller H.R. (1968). Superconductivity of small tin particles measured by tunneling. *Phys Rev Lett* 20: 1504–1507.

Girvin S.M., Glazman L.I., Jonson M., Penn D.R. and Stiles M.D. (1990). Quantum fluctuations and the single-junction Coulomb blockade. *Phys Rev Lett* 64: 3183–3186.

Gorter C.J. (1951). A possible explanation of the increase of the electrical resistance of thin metal films at low temperatures and small field strengths. *Physica* 17: 777–780.

Grabert H. and Devoret M.H. (1992). *Single Charge Tunneling*, Plenum, New York.

Grabert H., Ingold G.-L., Devoret M.H., Esteve D., Pothier H. and Urbina C. (1991). Single electron tunneling rates in multijunction circuits. *Z Phys B* 84: 143–155.

Grom G.F., Lockwood D.J., McCaffery J.P., Labbe N.J., Fauchet P.M., White Jr. B., Diener J., Kovalev D., Koch F. and Tsybeskov L. (2000). Ordering and self-organization in nanocrystalline silicon. *Nature* 407: 358–361.

Grüner G. (1988). The dynamics of charge-density waves. *Rev Mod Phys* 60: 1129–1181.

Guo L., Leobandung E. and Chou S.Y. (1997). A room-temperature silicon single-electron metal-oxide-semiconductor memory with nanoscale floating-gate and ultranarrow channel. *Appl Phys Lett* 70: 850–852.

Hanafi H., Tiwari S. and Khan I. (1996). Fast and long retention-time nano-crystal memory. *IEEE Trans Electron Devices* 43: 1553–1558.

Harmans C.J.P.M. (1992). Next electron, please. *Physics World* 5: 50–55.

Hasegawa H. and Kasai S. (2001). Hexagonal binary decision diagram quantum logic circuits using Schottky in-plane and wrap-gate control of GaAs and InGaAs nanowires. *Physica E* 11: 149–154.

He J., Durrani Z.A.K. and Ahmed H. (2004a). Two-way switch for binary decision diagram logic using silicon single-electron transistors. *Microelectronic Engineering* 73–74: 712–718.

He J., Durrani Z.A.K. and Ahmed H. (2004b). Universal three-way few-electron switch using silicon single-electron transistors. *Appl Phys Lett* 85: 308–310.

Heij C.P., Hadley P. and Mooij J.E. (2001). Single-electron inverter. *Appl Phys Lett* 78: 1140–1142.

Hinds B.J., Nishiguchi K., Dutta A., Yamanaka T., Hatanani S. and Oda S. (2000). Two-gate transistor for the study of Si/SiO$_2$ interface in silicon-on-insulator nano-channel and nanocrystalline Si memory device. *Jap J Appl Phys* 39: 4637–4641.

Hiramoto T., Ishikuro H., Saito K., Fujii T., Saraya T., Hashiguchi G. and Ikoma T. (1996). Fabrication of Si nanostructures for single electron device applications by anisotropic etching. *Jpn J Appl Phys* 35: 6664–6667.

Hofmann S., Ducati C., Neill R.J., Piscanec S., Ferrari A.C., Geng J., Dunin-Borkowsk R.E. and Robertson J. (2003). Gold catalyzed growth of silicon nanowires by plasma enhanced chemical vapor deposition. *J Appl Phys* 94: 6005–6012.

Horiguchi S., Nagase M., Shiraishi K., Kageshima H., Takahashi Y. and Murase K. (2001). Mechanism of potential profile formation in silicon single-electron transistors fabricated using pattern-dependent oxidation. *Jpn J Appl Phys* 40: L29–L32.

Hu G.Y. and O'Connell R.F. (1994). Exact solution for the charge soliton in a one-dimensional array of small tunnel junctions. *Phys Rev B* 49: 16773–16776.

Ingold G.L. and Nazarov Y.V. (1992). Charge tunneling rates in ultrasmall junctions in Grabert H. and Devoret M.H. (eds) *Single Charge Tunneling* Plenum, New York.

Ingold G.-L., Wyrowski P. and Grabert H. (1991). Effect of the electromagnetic environment on the single electron transistor. *Z Phys B* 85: 443–449.

International Technology Roadmap for Semiconductors (ITRS): 2005 Edition, JEITA English-Language Publications.

Irvine A.C., Durrani Z.A.K. and Ahmed H. (2000). A high-speed silicon-based few-electron memory with metal-oxide-semiconductor field-effect transistor gain element. *J Appl Phys* 87: 8594–8603.

Irvine A.C., Durrani Z.A.K., Ahmed H. and Biesemans S. (1998). Single-electron effects in heavily doped polycrystalline silicon nanowires. *Appl Phys Lett* 73: 1113–1115.

Ishii T., Yano K., Sano T., Mine T., Murai F. and Seki K. (1997a). A 3-D single-electron-memory cell structure with 2F^2 per bit. *Proc IEEE Int Electron Devices Meeting*: 924–926.

Ishii T., Yano K., Sano T., Mine T., Murai F. and Seki K. (1997b). Verify: key to the stable single-electron-memory operation. *Proc IEEE Int Electron Devices Meeting*: 171–174.

Ishikuro H. and Hiramoto T. (1997). Quantum mechanical effects in the silicon quantum dot in a single-electron transistor. *Appl Phys Lett* 71: 3691–3693.

Ishikuro H. and Hiramoto T. (1999). On the origin of tunneling barriers in silicon single electron and single hole transistors. *Appl Phys Lett* 74: 1126–1128.

Ishikuro H., Fujii T., Saraya T., Hashiguchi G., Hiramoto T. and Ikoma T. (1996). Coulomb blockade oscillations at room temperature in a Si quantum wire metal-oxide-semiconductor field-effect transistor fabricated by anisotropic etching on a silicon-on-insulator substrate. *Appl Phys Lett* 68: 3585–3587.

Iwamura H., Akazawa M. and Amemiya Y. (1998). Single-electron majority logic circuits. *IEICE Trans Electron* E81-C: 42–48.

Jalil M.B.A. and Wagner M. (1999). Image-soliton method applied to finite multiple tunnel junctions. *Phys Rev B* 59: 4626–4629.
Jalil M.B.A., Ahmed H. and Wagner M. (1998). Analysis of multiple-tunnel junctions and their application to bidirectional electron pumps. *J Appl Phys* 84: 4617–4624.
Kaizawa T., Oya T., Arita M., Takahashi Y. and Choi J.-B. (2006). Multifunctional device using nanodot array. *Jpn J Appl Phys* 45: 5317–5321.
Kamins T.I. and Pianetta P.A. (1980). MOSFETs in laser-recrystallized poly-silicon on quartz. *IEEE Electron Dev Lett* 1: 214–216.
Kamins T.I., (1971). Hall mobility in chemically deposited polycrystalline silicon. *J Appl Phys* 42: 4357–4365.
Kamiya T., Durrani Z.A.K. and Ahmed H. (2002). Control of grain-boundary tunneling barriers in polycrystalline silicon. *Appl Phys Lett* 81: 2388–2390.
Kamiya T., Durrani Z.A.K., Ahmed H., Sameshima T., Furuta Y., Mizuta H. and Lloyd N. (2003). Reduction of grain-boundary potential barrier height in polycrystalline silicon with hot H_2O-vapor annealing probed using point-contact devices. *J Vac Sci and Tech B* 21: 1000–1003.
Kamiya T., Nakahata K., Ro K., Fortmann C.M. and Shimuzu I. (1999). High rates and very low temperature fabrication of polycrystalline silicon from fluorinated source gas and their transport properties. *Mater Res Soc Symp Proc* 557: 513.
Kamiya T., Nakahata K., Tan Y.T., Durrani Z.A.K. and Shimuzu I. (2001). Growth, structure, and transport properties of thin (>10 nm) n-type microcrystalline silicon prepared on silicon oxide and its application to single-electron transistor. *J Appl Phys* 89: 6265–6271.
Kanemitsu Y. (1995). Light emission from porous silicon and related materials. *Phys. Rep* 263: 1–91.
Kanemitsu Y., Okamoto S., Otobe M. and Oda S. (1997). Photoluminescence mechanism in surface-oxidized silicon nanocrystals. *Phys Rev B* 55: R7375–7378.
Kapetanakis E., Normand P., Tsoukalas D. and Beltsios K. (2002). Room-temperature single-electron charging phenomena in large-area nanocrystal memory obtained by low-energy ion beam synthesis. *Appl Phys Lett* 80: 2794–2796.
Kapetanakis E., Normand P., Tsoukalas D., Beltsios K., Stoemenos J., Zhang S. and van den Berg J. (2000). Charge storage and interface states effects in Si-nanocrystal memory obtained using low-energy Si^+ implantation and annealing. *Appl Phys Lett* 77: 3450–3452.
Kasai S. and Hasegawa H. (2002). A single electron binary-decision-diagram quantum logic circuit based on Schottky wrap gate control of a GaAs nanowire hexagon. *IEEE Electron Dev Lett* 23: 446–448.
Kasai S., Amemiya Y. and Hasegawa H. (2000). GaAs Schottky wrap-gate binary-decision-diagram devices for realization of novel single electron logic architecture. *IEDM Tech Dig* 2000: 585–588.
Kastner, M. A. (1993). Artificial Atoms, *Physics Today*, 46: 24–31.

Kastner M.A., Field S.B., Meirav U., Scott-Thomas J.H.F., Antoniadis D.A. and Smith H.I. (1989). Kastner *et al.* reply. *Phys Rev Lett* 63: 1894–1894.

Katayama K., Mizuta H., Müller H.-O., Williams D. and Nakazato K. (1999). Design and analysis of high-speed random access memory with Coulomb blockade charge confinement. *IEEE Trans Electron Devices* 46: 2210–2216.

Keller M.W., Martinis J.M., Zimmerman N.M. and Steinbach A.H. (1996). Accuracy of electron counting using a 7-junction electron pump. *Appl Phys Lett* 69: 1804–1806.

Khalafalla M.A.H., Durrani Z.A.K. and Mizuta H. (2004). Coherent states in a coupled quantum dot nanocrystalline silicon transistor. *Appl Phys Lett* 85: 2262–2264.

Khalafalla M.A.H., Mizuta H. and Durrani Z.A.K. (2003). Switching of single-electron oscillations in dual-gated nanocrystalline silicon point-contact transistors. *IEEE Trans Nanotechnology* 2: 271–276.

Kiehl R.A. and Ohshima T. (1995). Bistable locking of single-electron tunneling elements for digital circuitry. *Appl Phys Lett* 67: 2494–2496.

Kim I., Han S., Han K., Lee J. and Shin H. (1999). Room temperature single electron effects in a Si nano-crystal memory. *IEEE Electron Device Lett* 20: 630–631.

Kitade T. and Nakajima A. (2004). Application of highly-doped Si single-electron transistors to an exclusive-NOR operation. *Jpn J Appl Phys* 43: L418–L420.

Kitade T., Ohkura K. and Nakajima A. (2005). Room-temperature operation of an exclusive-OR circuit using a highly doped Si single-electron transistor. *Appl Phys Lett* 86: 123118–123110.

Koester T., Goldschmidtboeing F., Hadam B., Stein J., Altmeyer S., Spangenberg B. and Kurz H. (1998). Direct patterning of single electron tunneling transistors by high resolution electron beam lithography on highly doped molecular beam epitaxy grown silicon films. *J Vac Sci Technol B* 16: 3804–3807.

Kohno H., Kikuo I. and Oto K. (2005). Electron transport in Si nanochains/nanowires. *J Electron Microsc* 54(Suppl. 1): i15–i19.

Korotkov A.N. (1995). Wireless single-electron logic biased by alternating electric field. *Appl Phys Lett* 67: 2412–2414.

Korotkov A.N. and Likharev K.K. (1998). Single-electron-parametron-based logic devices. *J Appl Phys* 84: 6114–6126.

Korotkov A.N., Chen R.H. and Likharev K.K. (1995). Possible performance of capacitively coupled single-electron transistors in digital circuits. *J Appl Phys* 78: 2520–2530.

Köster T., Hadam B., Hofmann K., Gondermann J.J., Stein J., Hu S., Altmeyer S., Spangenberg B. and Kurz H. (1997). Metal-oxide-semiconductor-compatible silicon based single electron transistor using bonded and etched back silicon on insulator material. *J Vac Sci Technol B* 15: 2836–2839.

Kouwenhoven L.P., Johnson A.T., van der Vaart N.C., Harmans C.J.P.M. and Foxon C.T. (1991c). Quantized current in a quantum-dot turnstile using oscillating tunnel barriers. *Phys Rev Lett* 67: 1626–1629.

Kouwenhoven L.P., Johnson A.T., van der Vaart N.C., van der Enden A., Harmans C.J.P.M. and Foxon C.T. (1991b). Quantized current in a quantum dot turnstile. *Z Phys B* 85: 381–388.

Kouwenhoven L.P., van der Vart N.C., Johnson A.T., Kool W., Harmans C.J.P.M., Williamson J.G., Staring A.A.M., and Foxon C.T., (1991a). Single electron charging effects in semiconductor quantum dots. *Z Phys B* 85: 367–373. See also: Kouwenhoven L.P., Vaart N.C. van der, Johnson A.T., Harmans C.J.P.M., Williamson J.G., Staring A.A.M., Foxon C.T., (1991) in Rossler, K. (ed) *Festkorperprobleme/Advances in Solid State Physics* 31: 329–340. Vieweg, Braunschweig.

Kowenhoven L.P., Marcus C.M., McEuen P.L., Tarucha S., Westervelt R.M., and Wingreen N.S., (1997). Electron transport in quantum dots in Sohn L.L., Kowenhoven L.P., and Schön G. (eds) *Mesoscopic Electron Transport* Kluwer, Dordrecht.

Krupenin V.A., Lotkhov S.V. and Presnov D.E. (1997). Instability of single-electron memory at low temperatures in Al/AlO$_x$/Al structures. *JETP* 84: 190–196.

Kulik I.O. and Shekhter R.I. (1975). Kinetic phenomena and charge discreteness in granular media. *Sov Phys JETP* 41: 308–316.

Kuzmin L.S. and Likharev K.K. (1987). Direct experimental observation of correlated discrete single-electron tunnelling. *JETP Lett* 45: 495–497.

Kuzmin L.S., Delsing P., Claeson T., and Likharev K.K. (1989). Single-electron charging effects in one-dimensional arrays of ultrasmall tunnel junctions. *Phys Rev Lett* 62: 2539–2542.

Lafarge P., Pothier H., Williams E.R., Esteve D., Urbina C. and Devoret M.H., (1991). Direct observation of macroscopic charge quantization. *Z Phys B* 85: 327–332.

Lambe J. and Jaklevic R.C. (1969). Charge-quantization studies using a tunnel capacitor. *Phys Rev Lett* 22: 1371–1375.

Lee B.T., Park J.W., Park K.S., Lee C.H., Paik S.W., Lee S.D., Choi J.B., Min K.S., Park J.S., Hahn S.Y., Park T.J., Shin H., Hong S.C., Lee K., Kwon H.C., Park S.I., Kim K.T. and Yoo K-H. (1998). Fabrication of a dual-gate-controlled Coulomb blockade transistor based on a silicon-on-insulator structure. *Semiconductor Sci Technol* 13: 1463–1467.

Lee P.A. and Stone A.D. (1985). Universal conductance fluctuations in metals. *Phys Rev Lett* 55: 1622–1625.

Lent C.S., Tougaw P.D. and Porod W. (1993a). Bistable saturation in coupled quantum dots for quantum cellular automata. *Appl Phys Lett* 62: 714–716.

Lent C.S., Tougaw P.D., Porod W. and Bernstein G.H. (1993b). Quantum cellular automata. *Nanotechnology* 4: 49–57.

Leobandung E., Guo L., Wang Y. and Chou S.Y. (1995a). Single hole quantum dot transistors in silicon. *Appl Phys Lett* 67: 2338–2340.

Leobandung E., Guo L., Wang Y. and Chou S.Y. (1995b). Observation of quantum effects and Coulomb blockade in silicon quantum-dot transistors at temperatures over 100 K. *Appl Phys Lett* 67: 938–940.

Leobandung E., Guo L., Wang Y. and Chou S.Y. (1995c). Single electron and hole quantum dot transistors operating above 110 K. *J Vac Sci Technol B* 13: 2865–2868.

Levinson J., Sheperd F.R., Scanlon P.J., Westwood W.D., Este G. and Rider M.J. (1982). Conductivity behavior in polycrystalline semiconductor thin film transistors. *J Appl Phys* 53: 1193–1202.

Likharev K.K. (1987). Single-electron transistors: Electrostatic analogs of the DC SQUIDs. *IEEE Trans Mag* 23: 1142–1145.

Likharev K.K. (1988). Correlated discrete transfer of single electrons in ultrasmall tunnel junctions. *IBM J Res Develop* 32: 144–158.

Likharev K.K. (1999). Single-electron devices and their applications. *Proceedings of the IEEE* 87: 606–632.

Likharev K.K. and Claeson T. (1992). Single electronics. *Scientific American* 266: 50–55.

Likharev K.K. and Korotkov A.N. (1995). Single-electron parametron. *Proc of IWCE-4*: p5.

Likharev K.K. and Zorin A.B. (1985). Theory of the Bloch-wave oscillations in small Josephson junctions. *J Low Temp Phys* 59: 347–382.

Likharev K.K., Bakhvalov N.S., Kazacha G.S. and Serdyukova S.I. (1989). Single-electron tunnel junction array: An electrostatic analog of the Josephson transmission line. *IEEE Trans Mag* 25: 1436–1439.

Littau K.A., Szajowski P.F., Muller A.J., Kortan A.R. and Brus L.E. (1993). A luminescent silicon nanocrystal colloid via a high-temperature aerosol reaction. *J Phys Chem* 97: 1224–1230.

Lotkhov S.V., Zangerle H., Zorin A.B. and Niemeyer J. (1999). Storage capabilities of a four-junction single-electron trap with an on-chip resistor. *Appl Phys Lett* 75: 2665–2667.

Martinis J.M., Nahum M. and Jensen H.D. (1994). Metrological accuracy of the electron pump. *Phys Rev Lett* 72: 904–907.

Matsumoto K., Gotoh Y., Maeda T., Dagata J.A. and Harris J.S. (2000). Room-temperature single-electron memory made by pulse-mode atomic force microscopy nano oxidation process on atomically flat alumina substrate. *Appl Phys Lett* 76: 239–241.

Matsuoka H. and Kimura S. (1995). Transport properties of a silicon single-electron transistor at 4.2 K. *Appl Phys Lett* 66: 613–615.

Matsuoka H., Ichiguchi T., Yoshimura T. and Takeda E. (1994). Coulomb blockade in the inversion layer of a Si metal-oxide-semiconductor field-effect transistor with a dual-gate structure. *Appl Phys Lett* 64: 586–588.

Matsuoka K.A., Likharev K.K., Dresselhaus P., Ji L., Han S. and Lukens J. (1997). Single-electron traps: A quantitative comparison of theory and experiment. *J Appl Phys* 81: 2269–2281.

McEuan P.L., Wingreen N.S., Foxman E.B., Kinaret J., Meirav U., Kastner M.A., Meir Y. and Wind S.J. (1993). Coulomb interactions and energy-level spectrum of a small electron gas. *Physica B* 189: 70–79.

Meir Y., Wingreen N.S. and Lee P.A. (1991). Transport through a strongly interacting electron system: Theory of periodic conductance oscillations. *Phys Rev Lett* 66: 3048–3051.

Meirav U. and Foxman E.B., (1996). Single-electron phenomena in semiconductors. *Semicond Sci Technol* 11: 255–284.

Meirav U., Kastner M.A. and Wind S.J. (1990). Single-electron charging and periodic conductance resonances in GaAs nanostructures. *Phys Rev Lett* 65: 771–774.

Mistry K. et al. (2007). A 45nm logic technology with high-k+ metal gate transistors, strained silicon, 9 Cu interconnect layers, 193nm dry patterning, and 100% Pb-free packaging. *IEDM* 2007: 247–250.

Miyoshi Y., Nakajima F., Motohisa J. and Fukui T. (2005). A 1 bit binary-decision-diagram adder circuit using single-electron transistors made by selective-area metalorganic vapor-phase epitaxy. *Appl Phys Lett* 87: 33501–33503.

Mizuta H., Müller H.-O., Tsukagoshi K., Williams D., Durrani Z., Irvine A., Evans G., Amakawa S., Nakazato K. and Ahmed H. (2001). Nanoscale Coulomb blockade memory and logic devices. *Nanotechnology* 12: 155–159.

Mizuta H., Wagner M. and Nakazato K. (2001). The role of tunnel barriers in phase-state low electron-number drive transistors (PLEDTRs). *IEEE Trans Electron Devices* 48: 1103–1108.

Müller H.-O, Williams D.A., Mizuta H. and Durrani Z.A.K. (2000). Simulating Si multiple tunnel junctions from pinch-off to ohmic conductance. *Material Science and Engineering B* 74: 36–39.

Müller H.-O., Williams D.A., Mizuta H., Durrani Z.A.K., Irvine A.C. and Ahmed H. (1999). Simulation of Si multiple tunnel junctions. *Physica B* 272: 85–87.

Nagamune Y., Sakaki H., Kowenhoven L.P., Mur L.C., Harmans C.J.P.M., Motohisa J. and Noge H. (1994). Single electron transport and current quantization in a novel quantum dot structure. *Appl Phys Lett* 64: 2379–2381.

Nakahata K., Ro K., Suemasu A., Kamiya T., Fortmann C.M. and Shimuzu I. (2000). Fabrication of polycrystalline silicon films from $SiF_4/H_2/SiH_4$ gas mixture using very high frequency plasma enhanced chemical vapor deposition with *in situ* plasma diagnostics and their structural properties. *Jpn J Appl Phys* 39: 3294–3301.

Nakajima A., Futatsugi T., Kosemura K., Fukano T. and Yokoyama N. (1997a). Single-electron traps: A quantitative comparison of theory and experiment. *Appl Phys Lett* 70: 1742–1744.

Nakajima A., Futatsugi T., Kosemura K., Fukano T. and Yokoyama N. (1997b). Si single electron tunneling transistor with nanoscale floating dot stacked on a Coulomb island by self-aligned process. *Appl Phys Lett* 71, 353–355.
Nakajima A., Futatsugi T., Kosemura K., Fukano T. and Yokoyama N. (1999a). Si single-electron tunneling transistor with nanoscale floating dot stacked on a Coulomb island by self-aligned process. *J Vac Sci Technol B* 17: 2163–2171.
Nakajima A., Ito Y. and Yokoyama S. (2002a). Conduction mechanism of Si single-electron transistor having a one-dimensional regular array of multiple tunnel junctions. *Appl Phys Lett* 81: 733–735.
Nakajima A., Sugita Y., Kawamura K., Tomita H. and Yokoyama N. (1996). Microstructure and optical absorption properties of Si nanocrystals fabricated with low-pressure chemical-vapor deposition. *J Appl Phys* 80: 4006–4011.
Nakajima F., Kumkura K., Motohisa J. and Fukui T. (1999b). GaAs single electron transistors fabricated by selective area metalorganic vapor phase epitaxy and their application to single electron logic circuits. *Jpn J Appl Phys* 38: 415–417.
Nakajima F., Miyoshi Y., Motohisa J. and Fukui T. (2003). Single-electron AND/NAND logic circuits based on a self-organized dot network. *Appl Phys Lett* 83: 2680–2682.
Nakajima F., Ogasawara Y., Motohisa J. and Fukui T. (2000b). Two-way current switch using Coulomb blockade in GaAs quantum dots by selective area metalorganic vapor phase epitaxy. *Physica E* 13: 703–707.
Nakajima F., Ogasawara Y., Motohisa J. and Fukui T. (2001). GaAs dot-wire coupled structures grown by selective area metalorganic vapor phase epitaxy and their application to single electron devices. *J Appl Phys* 90: 2606–2611.
Nakajima Y., Takahashi Y., Horiguchi S., Iwadate K., Namatsu H., Kurihara K. and Tabe M. (1994). Fabrication of a silicon quantum wire surrounded by silicon dioxide and its transport properties. *Appl Phys Lett* 65: 2833–2835.
Nakazato K., Blaikie R.J. and Ahmed H. (1994). Single-electron memory. *J Appl Phys* 75: 5123–5134.
Nakazato K., Blaikie R.J., Cleaver J.R.A. and Ahmed H. (1993). Single-electron memory. *Electronics Letters* 29: 384–385.
Nakazato K., Thornton T.J., White J. and Ahmed H., (1992). Single-electron effects in a point contact using side-gating in delta-doped layers. *Appl Phys Lett* 61: 3145–3147.
Namatsu H., Takahashi Y., Nagase M. and Murase K. (1995). Fabrication of thickness-controlled silicon nanowires and their characteristics. *J Vac Sci Technol B* 13: 2166–2169.
Namatsu H., Nagase M., Kurihara K., Horiguchi S. and Makino T. (1996). Fabrication of one-dimensional silicon nanowire structures with a self-aligned point contact. *Jpn J Appl Phys* 35: L1148–L1150.

Natori K., Uehara T. and Sano N. (2000), A Monte Carlo study of current–voltage characteristics of the scaled-down single-electron transistor with a silicon rectangular parallelepiped quantum dot. *Jpn J Appl Phys* 39: 2550–2555.

Nazarov Y.V. (1989). Coulomb blockade of tunneling in isolated junctions. *JETP Lett* 49: 126–128.

Neugebauer C.A. and Webb M.B. (1962). Electrical conduction mechanism in ultrathin, evaporated metal films. *J Appl Phys* 33: 74–82.

Nishiguchi K. and Oda S. (2000). Electron transport in a single silicon quantum structure using a vertical silicon probe. *J Appl Phys* 88: 4186–4190.

Normand P., Kapetanakis E., Dimitrakis P., Tsoukalas D., Beltsios K., Cherkashin N., Bonafos C., Benassayag G., Coffin H., Claverie A., Soncini V., Agarwal A. and Ameen M. (2003). Effect of annealing environment on the memory properties of thin oxides with embedded Si nanocrystals obtained by low-energy ion-beam synthesis. *Appl Phys Lett* 83: 168–170.

Oda S. and Otobe M. (1995). Preparation of nanocrystalline silicon by pulsed plasma processing. *Mater Res Soc Symp Proc* 358: 721–731.

Odintsov A.A. (1991). Single electron transport in a two-dimensional electron gas system with modulated barriers: A possible dc current standard. *Appl Phys Lett* 58: 2695–2697.

Ohata A., Toriumi A. and Uchida K. (1997). Coulomb blockade effects in edge quantum wire SOI MOSFETs. *Jpn J Appl Phys* 36: 1686–1689.

Ohshima T. (1996). Stability of binary logic tunneling phase states in dc-biased and ac-pumped single-electron tunnel junctions. *Appl Phys Lett* 69: 4059–4061.

Ohshima T. and Kiehl R.A. (1996). Operation of bistable phase-locked single-electron tunneling logic elements. *J Appl Phys* 80: 912–923.

Ono Y. and Takahashi Y. (2000). Single-electron pass-transistor logic: Operation of its elemental circuit. *IEDM Tech Dig* 297–300.

Ono Y. and Takahashi Y. (2003a). Electron pump by a combined single-electron/field-effect- transistor structure. *Appl Phys Lett* 82: 1221–1223.

Ono Y., Fujiwara A., Nishiguchi K., Inokawa H. and Takahashi Y. (2005). Manipulation and detection of single electrons for future information processing. *J Appl Phys* 97: 31101–31119.

Ono Y., Takahashi Y., Yamazaki K., Nagase M., Namatsu H., Kurihara K. and Murase K. (2000a). Fabrication method for IC-oriented Si single-electron transistors. *IEEE Trans Electron Devices* 47: 147–153.

Ono Y., Takahashi Y., Yamazaki K., Nagase M., Namatsu H., Kurihara K. and Murase K. (2000c). Si complementary single-electron inverter with voltage gain. *Appl Phys Lett* 76: 3121–3123.

Ono Y., Zimmerman N.M., Yamazaki K. and Takahashi Y. (2003b). Turnstile operation using a silicon dual-gate single-electron transistor. *Jpn J Appl Phys* 42, L1109–L1111.

Ono Y., Takahashi Y., Yamazaki K., Nagase M., Namatsu H., Kurihara K. and Murase K. (2000b). Single-electron transistor and current-switching device fabricated by vertical pattern-dependent oxidation. *Jpn J Appl Phys* 39: 2325–2328.

Orlov AO., Amlani I., Bernstein G.H., Lent C.S. and Snider G.L. (1997). Realization of a functional cell for quantum-dot cellular automata. *Science* 277: 928–930.

Otobe M., Kanai T. and Oda S. (1995). Fabrication of nanocrystalline Si by SiH_4 plasma cell. *Mater Res Soc Symp Proc* 377: 51–56.

Pace C., Crupi F., Lombardo S., Gerardi C. and Cocorullo G. (2005). Room-temperature single-electron effects in silicon nanocrystal memories. *Appl Phys Lett* 87: 182106–182108.

Park J.W., Park K.S., Lee B.T., Lee C.H., Lee S.D., Choi J.B., Yoo K-H., Kim J., Oh S.C., Park S.I., Kim K.T. and Kim J.J. (1999). Enhancement of Coulomb blockade and tunability by multidot coupling in a silicon-on-insulator-based single-electron transistor. *Appl Phys Lett* 75: 566–568.

Paul D.J., Cleaver J.R.A., Ahmed H. and Whall T.E. (1993). Coulomb blockade in silicon based structures at temperatures up to 50 K. *Appl Phys Lett* 63: 631–632.

Peng H.Y., Wang N.W., Shi W.S., Zhang Y.F., Lee C.S. and Lee S.T. (2001). Bulk-quantity Si nanosphere chains prepared from semi-infinite length Si nanowires. *J Appl Phys* 89: 727–731.

Peters M.G., den Hartog S.G., Dijkhuis J.I., Buyk O.J.A. and Molenkamp L.W. (1998). Single electron tunneling and suppression of short-channel effects in submicron silicon transistors. *J Appl Phys* 84: 5052–5056.

Pooley D.M., Ahmed H., Mizuta H. and Nakazato K. (1999). Coulomb blockade in silicon nano-pillars. *Appl Phys Lett* 74: 2191–2193.

Pothier H., Lafarge P., Orfila P.F., Urbina C., Esteve D. and Devoret M.H. (1991). Single electron pump fabricated with ultrasmall normal tunnel junctions. *Physica B* 169: 573–574.

Pothier H., Lafarge P., Urbina C., Esteve D. and Devoret M.H. (1992). Single-electron pump based on charging effects. *Europhys Lett* 17: 249–254.

Rafiq M.A., Durrani Z.A.K., Mizuta H., Colli A., Servati P., Ferrari A.C., Milne W.I. and Oda S. (2008). Room temperature single electron charging in single silicon nanochains. *J Appl Phys* 103: 53705–53708.

Rafiq M.A., Tsuchiya Y., Mizuta H., Oda S., Uno S., Durrani Z.A.K. and Milne W.I. (2006). Hopping transport in size-controlled Si nanocrystals. *J Appl Phys* 100: 14303–14306. See also: Rafiq M.A., Durrani Z.A.K., Mizuta H., Hassan M.M. and Oda S. (2008). Field-dependent hopping conduction in silicon nanocrystal films. *J Appl Phys* 104: 123710–123713.

Reed M. A. (1993). Quantum dots. *Scientific American* 268: 118–123.

Reed M.A., Randall J.N., Aggarwal R.J., Matyi R.J., Moore T.M. and Wetsel A.E., (1988). Observation of discrete electronic states in a zero-dimensional semiconductor nanostructure. *Phys Rev Lett* 60: 535–537.

Rokhinson L.P., Guo L.J., Chou S.Y. and Tsui D.C. (2000). Double-dot charge transport in Si single-electron/hole transistors. *Appl Phys Lett* 76: 1591–1593.

Saitoh M. and Hiramoto T. (2003). Room-temperature operation of highly functional single-electron transistor logic based on quantum mechanical effect in ultra-small silicon dot. *Proc IEEE Int Electron Devices Meeting*: 31.5.1–31.5.4.

Saitoh M. and Hiramoto T. (2004). Extension of Coulomb blockade region by quantum confinement in the ultrasmall silicon dot in a single-hole transistor at room temperature. *Appl Phys Lett* 84: 3172–3174.

Saitoh M., Harata H. and Hiramoto T. (2005). Room-temperature operation of current switching circuit using integrated silicon single-hole transistors. *Jpn J Appl Phys* 44: L338–L341.

Saitoh M., Saito T., Inukai T. and Hiramoto T. (2001a). Transport spectroscopy of the ultrasmall silicon quantum dot in a single-electron transistor. *Appl Phys Lett* 79: 2025–2027.

Saitoh M., Takahashi N., Ishikuro H. and Hiramoto T. (2001b). Large electron addition energy above 250 meV in a silicon quantum dot in a single-electron transistor. *Jpn J Appl Phys* 40: 2010–2012.

Sakamoto T., Kawaura H. and Baba T. (1999). Single-electron memory fabricated from doped silicon-on-insulator film. *Jpn J Appl Phys* 38: 5851–5852.

Sakamoto T., Kawura H. and Baba T. (1998). Single-electron transistors fabricated from a doped-Si film in a silicon-on-insulator substrate. *Appl Phys Lett* 72: 795–796.

Schön G. and Zaikin A.D. (1990). Quantum coherent effects, phase transitions, and the dissipative dynamics of ultra small tunnel junctions. *Physics Reports* 198: 237–412.

Schönenberger C., van Houten H. and Donkersloot H.C. (1992). Single-electron tunnelling observed at room temperature by scanning-tunnelling microscopy. *Europhys Lett* 20: 249–254.

Scott-Thomas J.H.F., Field S.B., Kastner M.A., Smith H.I. and Antoniadis D.A. (1989). Conductance oscillations periodic in the density of a one-dimensional electron gas. *Phys Rev Lett* 62: 583–586.

Scott-Thomas J.H.F., Kastner M.A., Antoniadis D.A., Smith H.I. and Field S. (1988). Si metal-oxide semiconductor field effect transistor with 70-nm slotted gates for study of quasi-one-dimensional quantum transport. *J Vac Sci Technol B* 6: 1841–1844.

Seto J.Y.W. (1975). The electrical properties of polycrystalline silicon films. *J Appl Phys* 46: 5247–5254.

Shklovskii B.I. and Efros A.L. (1984). *Electronic Properties of Doped Semiconductors* Springer, Berlin.

Shur M. (1987). *GaAs devices and circuits* Plenum Press, New York.

Smith R.A. and Ahmed H. (1997a). Gate controlled Coulomb blockade effects in the conduction of a silicon quantum wire. *J Appl Phys* 81: 2699–2703.

Smith R.A. and Ahmed H. (1997b). A silicon Coulomb blockade device with voltage gain. *Appl Phys Lett* 71: 3838–3840.
Special Issue on Single Charge Tunneling. (1991). Grabert H. (ed) *Z Physik B* 85.
Special Issue on Technology Challenges for Single Electron Devices. (1998). *IEICE Trans Electron* E81-C.
Special Issue on The Physics of Few-Electron Nanostructures. (1993). Geerligs L.J., Harmans C.J.P.M. and Kouwenhoven L.P. (eds) *Physica B* 189.
Stone N.J. and Ahmed H. (2000). Single-electron detector and counter. *Appl Phys Lett* 77: 744–746.
Stone N.J. and Ahmed H. (1998a). Silicon single-electron memory structure. *Microelectronic Engineering* 41/42: 511–514.
Stone N.J. and Ahmed H. (1998b). Silicon single electron memory cell. *Appl Phys Lett* 73: 2134–2136.
Stone N.J. and Ahmed H. (1999). Logic circuit elements using single-electron tunnelling transistors. *Electronics Letters* 35: 1883–1884.
Stuart B., Field M.A., Kastner U., Meirav U., Scott-Thomas J.H.F., Antoniadis D.A., Smith H.I. and Wind S.J. (1990). Conductance oscillations periodic in the density of one-dimensional electron gases. *Phys Rev B* 42: 3523–3536.
Sunamura H., Sakamoto T., Nakamura Y., Kawaura H., Tsai J S. and Baba T. (1999). Single-electron memory using carrier traps in a silicon nitride layer. *Appl Phys Lett* 74: 3555–3557.
Sze S.M. (2002). *Semiconductor devices, physics and technology* Wiley, New York.
Takahashi Y., Horiguchi S., Fujiwara A. and Murase K. (1996b). Co-tunneling current in very small Si single-electron transistors. *Physica B* 227: 105–108.
Takahashi N., Ishikuro H. and Hiramoto T. (2000a). Control of Coulomb blockade oscillations in silicon single electron transistors using silicon nanocrystal floating gates. *Appl Phys Lett* 76: 209–211.
Takahashi Y., Fujiwara A., Yamazaki K., Namatsu H., Kurihara K. and Murase K. (2000b). Multigate single-electron transistors and their application to an exclusive-OR gate. *Appl Phys Lett* 76: 637–639.
Takahashi Y., Fujiwara A., Yamazaki K., Namatsu H., Kurihara K. and Murase K. (1998). Si memory device operated with a small number of electrons by using a single-electron-transistor detector. *Electronics Letters* 34: 45–46.
Takahashi Y., Nagase M., Namatsu H., Kurihara K., Iwadate K., Nakajima Y., Horiguchi S., Murase K. and Tabe M. (1995). Fabrication technique for Si single-electron transistor operating at room temperature. *Electronics Letters* 31: 136–137.
Takahashi Y., Namatsu H., Kurihara K., Iwadate K., Nagase M. and Murase K. (1996a). Size dependence of the characteristics of Si single-electron transistors on SIMOX substrates. *IEEE Trans Electron Devices* 43: 1213–1217.
Takahashi Y., Ono Y., Fujiwara A. and Inokawa H. (2002). Silicon single-electron devices. *J Phys: Condens Matter* 14: R995–R1033.

Tan Y.T., Durrani Z.A.K. and Ahmed H. (2001a). Electrical and structural properties of solid phase crystallized polycrystalline silicon and their correlation to single-electron effects. *J Appl Phys* 89: 1262–1270.

Tan Y.T., Kamiya T., Durrani Z.A.K. and Ahmed H. (2001b). Single-electron effects in side-gated point contacts fabricated in low-temperature deposited nanocrystalline silicon films. *Appl Phys Lett* 78: 1083–1085.

Tan Y.T., Kamiy T., Durrani Z.A.K. and Ahmed H. (2003). Room temperature nanocrystalline silicon single-electron transistors. *J Appl Phys* 94: 633–637.

Taur Y. and Ning T.H. (1998). *Fundamentals of Modern VLSI Devices* Cambridge University Press, Cambridge.

Tilke A., Blick R.H., Lorenz H., Kotthaus J.P. and Wharam D.A. (1999). Coulomb blockade in quasimetallic silicon-on-insulator nanowires. *Appl Phys Lett* 75: 3704–3706.

Tiwari S., Hanafi H., Hartstein A., Crabbé E.F. and Chan K. (1996a). A silicon nanocrystals based memory. *Appl Phys Lett* 68: 1377–1379.

Tiwari S., Rana F., Chan K., Shi L. and Hanafi H. (1996b). Single charge and confinement effects in nano-crystal memories. *Appl Phys Lett* 69: 1232–1234.

Tougaw P.D. and Lent C.S. (1994). Logical devices implemented using quantum cellular automata. *J Appl Phys* 75: 1818–1825.

Tougaw P.D. and Lent C.S. (1996). Dynamic behavior of quantum cellular automata. *J Appl Phys* 80: 4722–4736.

Tougaw P.D., Lent C.S. and Porod W. (1993). Bistable saturation in coupled quantum-dot cells. *J Appl Phys* 74: 3558–3566.

Tringe J.W. and Plummer J.D. (2000). Electrical and structural properties of polycrystalline silicon. *J Appl Phys* 87: 7913–7926.

Tsukagoshi, K. and Nakazato, K. (1997). Electron pump current by two pulses with phase delay, *Appl Phys Lett* 71: 3138–3140.

Tsukagoshi K. and Nakazato K. (1998). Two-way switching based on turnstile operation. *Appl Phys Lett* 72: 1084–1085.

Tsukagoshi K., Alphenaar B.W. and Nakazato K. (1998). Operation of logic function in a Coulomb blockade device. *Appl Phys Lett* 73: 2515–2517.

Tsukagoshi K., Nakazato K., Ahmed H. and Gamo K. (1997). Electron pump in multiple-tunnel junctions. *Phys Rev B* 56: 3972–3975.

Tucker J.R. (1992). Complementary digital logic based on the "Coulomb blockade". *J Appl Phys* 72: 4399–4413.

Uchida K., Koga J., Ohba R. and Toriumi A. (2003). Programmable single-electron transistor logic for future low-power intelligent LSI: proposal and room-temperature operation. *IEEE Trans Electron Devices* 50: 1623–1630.

Uchida K., Koga J., Ohba R., Takagi S. and Toriumi A. (2001). Silicon single-electron tunneling device fabricated in an undulated ultrathin silicon-on-insulator film. *J Appl Phys* 90: 3551–3557.

Uchida K., Matsuzawa K. and Toriumi A. (1999). A new design scheme for logic circuits with single electron transistors. *Jpn J Appl Phys* 38: 4027–4032.

van der Wiel W.G., De Franceschi S., Elzerman J.M., Fujisawa T., Tarucha S., and Kowenhoven L.P. (2003). Electron transport through double quantum dots. *Rev Mod Phys* 75: 1–22.

van Houten H., Beenakker C.W.J. and Staring A.A.M. (1992). Coulomb-blockade oscillations in semiconductor nanostructures in Grabert H. and Devoret M.H. (eds) *Single Charge Tunneling* Plenum, New York.

van Houten H. and Beenaker C.W.J. (1989). Comment on "Conductance oscillations periodic in the density of a one-dimensional electron gas". *Phys Rev Lett* 63: 1893–1893.

van Wees B.J., van Houten H., Beenakker C.W.J., Williamson J.G., Kouwenhoven L.P., van der Marel D. and Foxon C.T. (1988). Quantized conductance of point contacts in a two-dimensional electron gas. *Phys Rev Lett* 60: 848–850.

Visscher E.H., Verbrugh S.M., Lindeman J., Hadley P. and Mooij J.E. (1995). Fabrication of multilayer single-electron tunneling devices. *Appl Phys Lett* 66: 305–307.

Volmar U.E., Weber U., Houbertz R. and Hartmann U. (1998). I(V) characteristics of one-dimensional tunnel junction arrangements. *Appl Phys A* 66: S735–S739.

Wagner R.S. and Ellis V.C. (1964). Vapor-liquid-solid mechanism of single crystal growth. *Appl Phys Lett* 4: 89–90.

Wasshuber C., Kosina H. and Selberherr S. (1997). SIMON – A simulator for single-electron tunnel devices and circuits. *IEEE Trans Computer-Aided Design of Integrated Circuits and Systems* 16: 937–944.

Welser J.J., Tiwari S., Rishton S., Lee K.Y. and Lee Y. (1997). Room temperature operation of a quantum-dot flash memory. *IEEE Electron Device Lett* 18: 278–280.

Wharam D.A. Thornton T.J., Newbury R., Pepper M., Ahmed H., Frost J.E.F., Hasko D.G., Peacock D.C., Ritchie D.A. and Jones G.A.C. (1988). One-dimensional transport and the quantisation of the ballistic resistance. *J Phys C* 21: L209–L214.

Williams E.R., Ghosh R.N. and Martinis J.M. (1992). Measuring the electron's charge and the fine-structure constant by counting electrons on a capacitor. *J Res Natl Inst Stand Techn* 97: 299–304.

Wilson W.L., Szajowski P.F. and Brus L.E. (1993). Quantum confinement in size-selected, surface-oxidized silicon nanocrystals. *Science* 262: 1242–1244.

Wu N-J., Asahi N. and Amemiya Y. (1997). Cellular-automaton circuits using single-electron-tunneling junctions. *Jpn J Appl Phys* 36: 2621–2627.

Wu W., Gu J., Ge H., Keimel C. and Chou S.Y. (2003). Room-temperature Si single-electron memory fabricated by nanoimprint lithography. *Appl Phys Lett* 83: 2268–2270.

Wu Y., Cui Y., Huynh L., Barrelet C.J., Bell D.C. and Lieber C.M. (2004). Controlled growth and structures of molecular-scale silicon nanowires. *Nano Lett* 4: 433–436.

Yamada Y., Kinoshita Y., Kasai S., Hasegawa H. and Amemiya Y. (2001). Quantum-dot logic circuits based on the shared binary-decision diagram. *Jpn J Appl Phys* 40: 4485–4488.

Yano K., Sasaki Y., Rikino K. and Seki K. (1996b). Top-down pass-transistor logic design. *IEEE J. Solid-State Circuits* 31: 792–803.

Yano K., Ishii T., Hashimoto T., Kobayashi T., Murai F. and Seki K. (1993). A room-temperature single-electron memory device using fine-grain polycrystalline silicon. *Proc IEEE Int Electron Devices Meeting*: 541–544.

Yano K., Ishii T., Hashimoto T., Kobayashi T., Murai F. and Seki K. (1994). Room-temperature single-electron memory. *IEEE Trans Electron Devices* 41: 1628–1638.

Yano K., Ishii T., Sano T., Hashimoto T., Kobayashi T., Murai F. and Seki K. (1995). Transport characteristics of polycrystalline-silicon wire influenced by single-electron charging at room temperature. *Appl Phys Lett* 67: 828–830.

Yano, K., Ishii, T., Sano, T., Mine, T., Murai, F. and Seki, K. (1996a). Single-electron-memory integrated circuit for giga-to-tera bit storage, *Proc. IEEE Int. Solid-State Circuits Conf.* pp. 266–267.

Yano K., Ishii T., Sano T., Mine T., Murai F., Hashimoto T., Kobayashi T., Kure T. and Seki K. (1999). Single-electron memory for giga-to-tera bit storage. *Proceedings of the IEEE* 87: 633–651.

Yano K., Ishii T., Sano T., Mine T., Murai F., Kure T. and Seki K. (1998). A 128 Mb early prototype for gigascale single-electron memories. *Proc IEEE Int Solid-State Circuits Conf*: 344–345.

Yoo K-H., Park J. W., Kim J., Park K.S., Oh S.C., Lee J-O., Kim J.J., Choi J.B. and Lee J.J. (1999). An in-plane GaAs single-electron memory cell operating at 77 K. *Appl Phys Lett* 74: 2073–2075.

Zeller H.R. and Giaever I, (1969). Tunneling, zero-bias anomalies, and small superconductors. *Phys Rev* 181: 789–799.

Zhong Z., Fang Y., Lu W. and Lieber C.M. (2005). Coherent single charge transport in molecular-scale silicon nanowires. *Nano Lett* 5: 1143–1146.

Zhong Z., Wang D., Cui Y., Bockrath M.W. and Lieber C.M. (2003). Nanowire crossbar arrays as address decoders for integrated nanosystems. *Science* 302: 1377–1379.

Zhuang L., Guo L. and Chou S.Y. (1998). Silicon single-electron quantum-dot transistor switch operating at room temperature. *Appl Phys Lett* 72: 1205–1207.

Zimmerli G., Kautz R.L. and Martinis J.M. (1992). Voltage gain in the single-electron transistor. *Appl Phys Lett* 61: 2616–2618.

Zwerger W. and Scharpf M. (1991). Crossover from Coulomb-blockade to ohmic conduction in small tunnel junctions. *Z Phys B* 85: 421–426.

Index

2-DEG, 11, 20, 59, 61, 85, 103, 119, 175, 181, 182
activation energy, 9, 105, 110, 113
addition energy, of quantum dot, 19, 63
alignment marks, 96, 127, 196
amorphous silicon, 108, 112, 113, 120, 143, 151
AND operation, 223, 227, 229, 232, 237, 238, 240, 243, 244, 245, 254, 255
anisotropic etching, 88
Arrhenius plot, 105, 110, 112, 114
atomic force microscope (AFM), 132, 133
background charge, 20, 140, 144, 145, 146, 227, 228
BDD, 21, 228, 229, 230, 231, 232, 233, 234, 235, 236, 237, 238, 240, 241, 243, 245, 246, 251
binary decision diagram, 21, 175, 186, 191, see also 'BDD'
buried oxide layer (BOX), 77, 79, 82, 84, 93, 95, 96, 151, 193, 195, 196, 247
capacitance standard, 206
charge density wave, see also Wigner lattice, 72, 73
charge fluctuations, 3, 27, 140
charge packet, 4, 20, 21, 174, 209, 228, 231, 238, 239, 241
charge state logic, 21, 210, 228
charge-coupled devices, 21, 174, 175, 200, 201

circuit tree, in BDD logic, 231, 236, 241
complementary SET (C-SET), 211, 213, 215, 217, 218, 219, 220, 221, 222
co-tunnelling, 24, 208, 219
Coulomb blockade, 1, 7, 13, 14, 16, 17, 18, 20, 23, 27, 28, 30, 31, 34, 35, 38, 41, 42, 45, 46, 47, 49, 52, 63, 64, 65, 89, 106, 117, 122, 128, 130, 131, 134, 135, 149, 150, 156, 158, 168, 177, 180, 183, 185, 188, 197, 198, 199, 211, 212, 213, 214, 215, 216, 217, 225, 234, 238, 241, 242, 249
Coulomb diamonds, 56, 80, 90, 100, 101
Coulomb gap, 7, 8, 10, 11, 14, 15, 16, 46, 52, 53, 54, 59, 75, 79, 82, 84, 89, 91, 99, 100, 110, 114, 135, 137, 150, 153, 154, 155, 160, 161, 162, 164, 165, 170, 172, 173, 178, 179, 181, 185, 190, 191, 199, 208, 248, 249, 252
Coulomb oscillations, see also 'single-electron current oscillations', 54, 62, 64, 73, 80, 82, 84, 91, 92, 93, 119, 120, 121, 122, 123, 225
Coulomb staircase, 8, 10, 11, 18, 38, 52, 54, 59, 71, 82, 99, 109, 125, 127, 128, 129, 154
coupled quantum dots, 12, 179
critical charge, 18, 20, 27, 29, 30, 31, 35, 132, 134, 177

crystalline silicon, 19, 74, 75, 87, 101, 107, 108, 112, 113, 118, 124, 146, 153, 154, 155, 173, 238
current plateaus, 178, 184, 193, 195, 205
current standard, 206
defect states, 75, 77, 78, 105, 109, 155, 156, 228
delocalization, 24, 106, 117, 124
disorder, in material, 11, 74, 75, 88, 90, 107, 108, 110, 113, 118, 119, 238
double tunnel junction, 5, 10, 13, 14, 15, 18, 38, 46, 49, 52, 57, 65, 157, 186, 206, 211
DRAM, 2, 150, 164
effective capacitance, in MTJ, 66, 67, 129
electrometer, 11, 130, 132, 134, 135, 136, 137, 138, 144, 206, 207, 258, 260
electron pump, 13, 17, 20, 21, 174, 175, 179, 183, 184, 185, 186, 189, 190, 191, 193, 194, 198, 200, 203, 204, 205, 206, 208, 229, 238, 240, 246, 247, 249, 250, 251, 252
electron-cyclotron resonance plasma etching, 82
electron-cyclotron resonance plasma oxidation, 82
electrostatic coupling, 13, 119, 120, 121, 122, 252
electrostatic energy, 18, 26, 30, 31, 32, 33, 34, 38, 39, 40, 41, 49, 63, 178
energy diagram
  quantum dot, 63
  SET, 56, 57
  single-electron box, 36, 37
fan-in, 230
fan-out, 211, 230, 233, 254
Fermi energy, 14, 15, 25, 36, 56, 57, 60, 63, 72, 97, 101, 104, 122, 182, 258
Fermi level, see also 'Fermi energy', 13, 103, 155, 258
few-electron transfer, 20, 174, 175
FLASH memory, 2, 140, 141, 142

floating gate, 20, 140, 141, 143, 144, 224
Fowler-Nordheim (FN) tunnelling, 142, 145
gain-cell, 116, 138
gate
  back-gate, 77, 78, 107, 125, 127, 145
  central gate, 163, 164, 165, 177, 191, 192
  dual-gate, 72, 86, 87, 150
  outer gate, 150, 151, 163, 164, 165, 168, 170
  side-gate, 77, 78, 79, 81, 88, 90, 91, 93, 94, 96, 99, 107, 108, 110, 112, 120, 133, 134, 137, 139, 140, 150, 152, 153, 154, 155, 160, 162, 164, 165, 168, 172, 191, 193, 194, 195, 196, 219, 238, 240, 246, 248
  split-gate, 72, 150, 151, 163, 164, 165, 167, 168, 170
  top-gate, 77, 78, 80, 83, 84, 85, 88, 107, 117
  upper gate, 72, 86, 87, 201, 202
'global' view, 32
grain boundary (GB), 19, 101, 103, 104, 105, 106, 107, 108, 109, 110, 111, 112, 113, 114, 115, 116, 117, 118, 119, 120, 123, 124, 155, 156
grain-boundary engineering, 116
granular metal film, 8, 9, 10, 11, 23
heterostructure, 3, 11, 17, 59, 85, 181, 244
hexagonal network, of SETs, 238, 241, 243, 244
hysteresis, 20, 35, 135, 137, 140, 142, 144, 150, 160, 162, 165, 166, 168
universal hysteresis, 135
inverter, 21, 211, 213, 215, 216, 217, 218, 219, 220, 221, 222, 224, 232, 254
lithography
  electron-beam, 11, 12, 19, 74, 75, 78, 79, 80, 81, 82, 90, 92, 93, 95, 96, 108, 110, 112, 125, 126, 127,

136, 138, 143, 151, 152, 159, 163, 194, 196, 243, 247, 260
  nano-imprint, 143
  nanolithography, 23
  optical, 78, 93, 95, 97, 195, 196, 248
localization, 8, 24, 106, 117
majority logic, using SETs, 21, 227, 254
material synthesis techniques, 124
mean free path, of electrons, 61, 85
memory
  cell selection, 151, 168, 170
  data line, 138, 148, 149, 157
  erase operation, 141, 148, 149
  memory array, 138, 163, 164
  read operation, 144, 146, 148, 149, 166, 167, 168, 170
  word-line, 146, 147, 148, 149, 157, 158, 160, 162, 164, 165, 168, 169, 170, 171
  write operation, 138, 141, 145, 148, 149, 157, 158, 167, 168, 169, 170, 171
memory cell, 1, 2, 13, 17, 20, 36, 132, 133, 134, 135, 136, 137, 138, 140, 141, 142, 143, 145, 147, 148, 149, 150, 157, 160, 162, 165, 167, 168, 172, 229, 256
memory node, 20, 130, 134, 135, 136, 137, 138, 139, 140, 144, 145, 146, 147, 149, 150, 151, 156, 157, 158, 159, 161, 162, 163, 164, 165, 167, 168, 170, 171, 172, 222
  charge, 132, 135, 138, 146, 158
  capacitor, 136
  voltage, 134, 137, 158, 161, 170
mesa, 88, 93, 94, 95, 96, 97, 181, 195, 196
mesoscopic, 3, 72, 85
messenger, in BDD logic, 231, 232, 233, 234, 235, 236, 241, 245
metrological applications, 21, 174, 175, 206, 208
metrology, 11
milli-Kelvin, 23, 61, 85, 119, 132, 139, 181, 191, 193

minimum feature size, 1, 155
MOSFET, 1, 2, 4, 8, 11, 19, 20, 21, 72, 74, 75, 77, 84, 85, 86, 87, 88, 116, 130, 138, 141, 142, 143, 149, 150, 151, 156, 157, 160, 161, 162, 163, 164, 165, 166, 168, 170, 172, 174, 175, 200, 201, 202, 203, 204, 205, 210, 211, 214, 215, 220, 221, 224, 230, 234
MTJ, 12, 13, 15, 16, 17, 19, 20, 21, 64, 65, 66, 67, 68, 69, 70, 71, 81, 87, 90, 91, 93, 99, 101, 107, 109, 112, 124, 125, 128, 129, 132, 133, 134, 135, 136, 137, 138, 139, 144, 150, 153, 154, 155, 157, 158, 174, 175, 178, 184, 185, 186, 187, 188, 189, 190, 191, 192, 193, 194, 195, 198, 199, 200, 206, 221, 222, 226, 238, 248, 260
  offset voltage, 71
multiple-tunnel junction, see also MTJ, 12, 15, 36, 49, 64, 81, 132, 174, 221
NAND operation, 21, 219, 220, 221, 222, 232, 235, 236, 240, 244, 245, 259
nanochain, 20, 75, 124, 125, 126, 127, 128
nanocrystal, 19, 20, 71, 74, 101, 102, 103, 115, 117, 118, 119, 124, 125, 126, 127, 128, 129, 140, 141, 142, 143, 224, 238
nanocrystalline silicon, 19, 20, 74, 75, 101, 103, 107, 111, 117, 119, 130, 146, 156, 173
  nanocrystalline silicon (nc-Si) SET, 20, 110, 111, 115, 118, 130
nanoparticle, 9, 10
nanowire
  in silicon, 19, 20, 74, 75, 80, 82, 83, 84, 85, 87, 88, 89, 90, 91, 93, 96, 97, 98, 99, 107, 108, 109, 110, 111, 112, 116, 117, 118, 124, 125, 134, 136, 137, 139, 143, 144, 149, 150, 151, 152, 153, 154, 155, 156, 159, 162, 166, 173, 184, 193, 194, 195, 196,

201, 222, 242, 245, 246, 247, 248, 250, 251, 260
SET, 19, 75, 87, 90, 91, 93, 97, 99, 107, 108, 110, 112, 118, 134, 136, 137, 149, 150, 151, 153, 154, 155, 156, 173, 193, 195, 222, 237, 240, 245, 246, 250, 251
NOR operation, 21, 219, 220, 221, 222, 223, 232, 259
offset charge, 48, 49, 54, 56, 78, 140, 144, 146, 155, 177, 206, 211, 212, 218, 221, 227, 228
one-dimensional, 11, 60, 61, 64, 71, 72, 73, 74, 103, 124, 152
orthodox theory, 24, 34, 158
oxidation, 12, 19, 74, 75, 78, 82, 84, 85, 91, 92, 93, 96, 97, 99, 107, 108, 116, 124, 139, 143, 153, 154, 155, 194, 219, 247, 248
  selective, 112, 113, 114, 119, 120
parametron, single-electron, 21, 229, 258, 259, 260
pattern-dependent oxidation (PADOX), 74, 82, 83, 84, 85, 91, 98, 118, 203, 219, 226, 245
peak-to-valley ratio (PVR), 19, 75, 91, 92, 93, 114, 224, 226, 227, 246
percolation path, 105, 116, 146, 147
point-contact SET, 19, 89, 92, 107, 111, 112, 115, 118, 119, 120, 122, 123, 156, 173
polycrystalline (poly) silicon, 80, 83, 87, 88, 103, 106, 107, 108, 109, 110, 111, 115, 116, 118, 143, 151, 152, 154, 155, 156, 201
potential fluctuations, 87, 91, 191, 194
programmable logic, 211, 222, 223, 224
quantization, see also 'quantum confinement', 2, 10, 25, 92, 195
quantum cellular automaton (QCA), 21, 228, 229, 252, 253, 254, 255, 256, 257, 258
quantum confinement, 3, 11, 18, 19, 59, 63, 64, 84, 85, 86, 88, 89, 90, 91, 92, 98, 101, 102, 103, 107, 140, 146, 182
quantum dashes, 254
quantum dot (QD), 3, 4, 11, 12, 17, 18, 19, 21, 25, 59, 60, 61, 62, 63, 64, 85, 87, 89, 92, 101, 102, 103, 106, 107, 115, 119, 120, 123, 125, 175, 179, 181, 182, 183, 184, 229, 245, 252, 253, 254, 256
quantum fluctuation of charge, 27
quantum of resistance, 24, 190
quasi-molecular states, 119, 123, 124
radio frequency (r.f.) signal, 17, 20, 21, 174, 175, 176, 181, 183, 184, 185, 186, 187, 189, 190, 191, 192, 193, 194, 195, 196, 198, 199, 206, 239, 244, 245, 246, 249, 250, 252
random access memory, 2, 116, 150
reactive-ion etching (RIE), 75, 77, 78, 79, 80, 81, 90, 93, 95, 96, 108, 112, 136, 137, 159, 163, 167, 195, 196, 248
registration marks, 93, 94, 95, 96
resistively-loaded, 209, 211, 216, 218, 227
resonant tunnelling, 25, 57, 82, 90, 93, 106, 115
retention time, of memory, 142, 149, 158, 172, 173
room temperature operation, of SETs, 19, 71, 80, 82, 84, 88, 89, 91, 92, 102, 111, 112, 114, 155, 156, 226, 237, 244, 256
room temperature operation, of single-electron memory, 132, 140, 144, 142, 146, 155
scanning tunnelling microscope, 11
Schottky, 103, 104, 129, 240, 242, 243
self-limiting oxidation, 82, 107
sense amplifier, 130, 138, 146
SET, for specific types, see:
  nanocrystalline silicon (nc-Si) SET,
  nanowire SET,
  point-contact SET,
  see also 'single-electron transistor'

silicon grains, 101, 102, 106, 109, 112, 113, 117, 146, 152
silicon-on-insulator (SOI), 11, 12, 19, 21, 74, 75, 76, 77, 78, 79, 80, 81, 87, 88, 90, 92, 93, 99, 107, 117, 126, 136, 138, 143, 149, 150, 151, 152, 156, 159, 163, 167, 175, 193, 195, 201, 203, 219, 222, 223, 242, 245, 246, 247, 260
single tunnel junction, 18, 25, 26, 27, 28, 29, 31, 36, 132, 134
single-electron
   box, 10, 12, 13, 14, 15, 16, 18, 28, 31, 32, 35, 36, 38, 41, 56, 131, 132, 134
   charging energy, 6, 7, 8, 9, 13, 14, 23, 65, 73, 102, 125, 129, 148, 177, 185
   current oscillations, 14, 18, 88, 90, 91, 98, 100, 109, 112, 114, 150, 183
   detector, 138
   latch, 213, 236
   logic, 12, 21, 174, 209, 210, 221, 228, 229, 245
   memory, 1, 12, 13, 17, 20, 36, 130, 131, 132, 133, 138, 144, 146, 149, 174, 238, see also memory
   memory, background charge insensitive, 140, 144, 145
   memory, lateral single-electron (L-SEM), 138, 149, 150, 151, 156, 157, 159, 160, 162, 163, 164, 165, 167, 172, 173
   memory, temperature dependence, 172
   Monte Carlo simulation, 52, 54, 128, 71, 195
   pump, 13, 17, 20, 21, 174, 175, 179, 181, 183, 184, 190, 191, 203, 204, 206, 208
   transistor (general), see also 'SET' for specific types, 1, 3, 10, 11, 12, 13, 14, 16, 17, 18, 19, 20, 21, 23, 25, 37, 38, 46, 48, 49, 52, 54, 55, 56, 57, 60, 63, 64, 69, 71, 72, 74, 130, 210

trap, 36
tunnelling oscillations, 18, 27, 69, 70, 228
turnstile, 13, 17, 20, 21, 174, 175, 176, 177, 178, 179, 181, 183, 184, 185, 201, 203, 205, 208, 237, 238, 239, 240
single-hole transistor, 75, 76, 93, 224, 226, 246
soliton, in charge, 65, 68, 69, 70
stability diagram, 46, 48, 49, 51, 56, 59, 179, 186, 189, 195, 199, 202, 205, 212, 213, 214, 215, 216, 258
statistical fluctuations, in charge, 1
storage charge, see also memory node charge, 130
thermally activated current, 7, 8, 99, 172, 190
thermionic emission, 105, 106, 113
threshold voltage, 2, 69, 72, 97, 106, 128, 141, 142, 143, 144, 145, 146, 147, 168, 178
trench isolation, 76, 107, 136
tunnelling rates, 8, 18, 34, 38, 43, 44, 45, 46, 47, 48, 49, 50, 52, 53, 65, 178, 188
two-dimensional electron gas (2-DEG), 11, 20, 59, 61, 85, 103, 119, 175, 181, 182
two-way switch, 21, 228, 229, 230, 231, 238, 241, 243, 244, 245, 246, 249, 250, 251
universal conductance fluctuations, 72
vertical transport, of electrons, 115
voltage gain, 91, 211, 213, 219, 222, 226, 230
voltage state logic, 21, 209, 210, 211
wavefunction, 8, 20, 24, 64, 115, 119, 120, 121, 122, 123, 124
wavefunction coupling, 115, 123
Wigner lattice, 70, 72
wireless, single-electron logic, 21, 229, 259
XOR operation, 21, 225, 226, 227, 229, 232, 236, 238, 243, 255